# Zoophysiology Volume 11

Coordinating Editor: D. S. Farner

Editors:

W. S. Hoar B. Hoelldobler

K. Johansen H. Langer M. Lindauer

Miles H. A. Keenleyside

# Diversity and Adaptation in Fish Behaviour

With 67 Figures

Springer-Verlag
Berlin  Heidelberg  New York 1979

Professor M. H. A. KEENLEYSIDE
University of Western Ontario
Department of Zoology
London, Canada N6A 5B7

For explanation of the cover motive see legend to Fig. 4.5 (p. 73).

ISBN 3-540-09587-X Springer-Verlag Berlin Heidelberg New York
ISBN 0-387-09587-X Springer-Verlag New York Heidelberg Berlin

Library of Congress Cataloging in Publication Data. Keenleyside, Miles H A 1929—.
Diversity and adaptation in fish behavior. (Zoophysiology; v. 11) Bibliography: p. Includes
index. 1. Fishes—Behavior. I. Title. QL639.3.K43. 597′.05.79-17435.

© by Springer-Verlag Berlin Heidelberg 1979.
Printed in Germany.

Typesetting, printing and bookbinding: Brühlsche Universitätsdruckerei, Gießen.
2131/3130-543210

*This book is dedicated to
my wife Hilda and our sons
Eric and Joel.*

# Preface

Fish are extremely successful vertebrates. Because of a long and divergent history they are now found in almost every conceivable aquatic habitat. This radiation has been accompanied by great diversity in structure, physiology, and behaviour. Despite this variability, fish must solve a number of basic problems that are common to all animals. The most important of these are: (1) to find and ingest appropriate food; (2) to avoid predation; and (3) to reproduce. The main purpose of this book is to describe the variety of behavioural strategies that fish use in coping with these problems.

My approach has been to draw together material from both field and laboratory work that is widely scattered in the literature. The major emphasis is on field studies, since my main concern is with adaptive solutions to problems, and I believe these are most likely to be correctly perceived by workers who are familiar with the natural ecological setting of their animals. Of course, many details of behaviour cannot be seen and quantified adequately in the field, and therefore I have not ignored laboratory studies. However, even here I have concentrated on work that illustrates the variety of solutions that fish use to solve the three basic problems. Much important work, for instance on causation, development, and learning, is not included.

The number and quality of field studies in fish behaviour have increased greatly in recent years. This is largely due to advances in underwater technology, especially in the areas of free-diving gear, photographic and television equipment, small submersible craft, and self-contained benthic habitats (e.g., Clarke et al., 1967; Brock and Chamberlain, 1968; Myrberg, 1973). Also, field work need no longer be purely observational and descriptive. A recent article by the experienced field ethologist Hans Fricke (1977) illustrates the range of possibilities for the ethologist who takes experimental equipment with him when he dives. To be sure, field research can be expensive, and sometimes dangerous, but fish behaviourists can no longer claim that their study animals are totally inaccessible unless brought into the laboratory. In addition, the excitement and satisfaction derived from studying undisturbed, free-living animals are potent motivating forces for ethologists who are also naturalists.

Students of insect, bird, and mammal behaviour have known this for a long time. They are now being joined by fish behaviourists, as the aquatic environment becomes increasingly accessible.

One final point; the book is not an exhaustive review. It is a sampling of the pertinent literature, that is itself strongly biased towards a few groups of fish that ethologists have tended to concentrate on. It is selective, with examples chosen to illustrate the range and variety of behavioural adaptations that enable fishes to cope with the problems of food, predation, and reproduction.

London, Canada                                    M. H. A. KEENLEYSIDE
September, 1979

# Acknowledgements

The completion of this book required the assistance of many people. In particular I am indebted to William S. Hoar, who urged me to write it, and who provided much useful advice throughout its preparation.

The bulk of the writing was done during a one-year study leave at the University of British Columbia. I am grateful to that institution, and in particular to G.G.E. Scudder, Head of the Department of Zoology, for providing me with space and uninterrupted time for writing during that year. I thank the University of Western Ontario for granting me the study leave.

The staff of the Natural Sciences Library, UWO, and the Woodward Biomedical Library, UBC, were consistently helpful in finding materials for me. Hilda Keenleyside relieved me of many hours of work by translating some critical papers from German. Parts of the manuscript were typed by Helen Kyle, Karen Carsh, Jane Sexsmith, and Leslie Borleski. I am indebted to all these people for their help.

All the illustrations were prepared by Catherine Farley. Most have been slightly modified from the originals, and I am extremely grateful for the care and consistency with which she did this work.

The following publishers gave permission to use previously published illustrations: Academic Press Inc. (London) Ltd.; Angus and Robertson Publishers; Baillière Tindall; Ernest Benn Limited; E.J. Brill; Cambridge University Press; Fisheries and Environment Canada; Hutchinson Publishing Group Ltd.; Longman Group Limited; Macmillan Journals Ltd.; Marine Biological Laboratory, Woods Hole, Massachusetts; McGraw-Hill Book Co.; Methuen and Co. Ltd.; Munksgaard International Booksellers and Publishers; National Marine Fisheries Service (U.S.A.); National Research Council of Canada; Oxford University Press; Paul Parey; Springer-Verlag; Universitetsforlaget Oslo; University of Michigan Press; Weidenfeld (Publishers) Limited; John Wiley & Sons, Inc.; Wistar Institute Press; The Zoological Society of London.

# Contents

Chapter 7
**Social Organization** . . . . . . . . . . . . . . . . . 149

Chapter 1

# Locomotion

Behaviour is anything an animal does. More technically, it is "the whole complex of observable, recordable, or measurable activities of a living animal" (Verplanck, 1957, p. 2). Thus, we label as behaviour any activities of animals that we, or our instruments, detect. These activities usually involve overt movements of the animal or some of its parts. Some classes of behaviour, such as resting, colour pattern changes and sound production, require little or no movement for their expression. Nevertheless, most behaviour that has been studied involves movement, and therefore knowledge of the variety of forms that locomotion can take, and of the forces regulating locomotion, is important for an understanding of fish behaviour. The subject has recently been reviewed at length by Lindsey (1978) and is given briefer treatment here.

The most basic component in the locomotor machinery of fishes is the energy source, located in the muscles. The size, shape, and distribution throughout the body of the muscle packages vary greatly, and this in turn is largely responsible for the great variety in fish form, size, and locomotor capacity. Although the most efficient swimmers tend to have a fusiform or spindle-shaped body, and to be well streamlined, there are many other body types (Fig. 1.1). This plasticity of body shape illustrates the fact that fishes as a group are highly successful; they have adaptively radiated into most aquatic environments, and some are also capable of brief sojourns into air and onto land.

Despite this variety in body form, all fishes must solve the same two basic locomotor problems: (1) that of moving forward, or of holding position in flowing water, relative to a stationary background, and (2) that of slowing, stopping, turning, rising, and sinking. These can be called respectively the problems of *propulsion* and of *manoeuvering and stabilizing*. In general, both are solved by mechanisms that "...can be expected to evolve so as to decrease the energy required for swimming, to increase the maximum acceleration and speed of the fish and to improve manoeuvrability" (Alexander, 1967, p. 19).

## 1.1 Propulsion

The ease with which a fish moves through water is a function of several factors:

(a) The density of its body in relation to the water's density. Fish with no swimbladder, such as sharks, are negatively buoyant and must keep swimming to avoid sinking. Many teleosts have a swimbladder, by which they achieve neutral buoyancy. Those whose swimbladder opens to the exterior via a duct

1

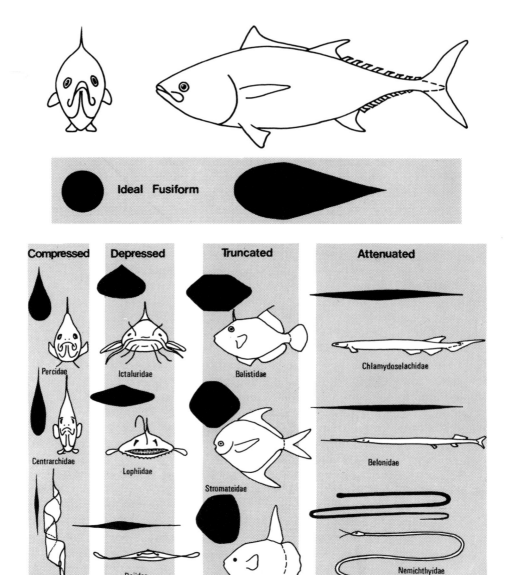

**Fig. 1.1.** Variation in body shape among fishes. (After Lagler et al., 1962)

(physostomes) are able to change their buoyancy by intake of air at the surface or release of swimbladder gas. Those with a closed bladder (physoclists) adjust their buoyancy much more slowly, by transfer of gas to or from the blood (Alexander, 1967).

(b) The speed and turbulence of the water. Natural waters are seldom still, except at great depths in the oceans, or in small bodies of water in calm weather.

2

Most fishes live in water that is kept in motion by wind, tides, or gravity. Even slight movements of its own body or fins set the water surrounding a fish in motion. Thus, few fish spend much time in absolutely still water (Gray, 1953).

(c) The effectiveness of its drag-reducing mechanisms (discussed below in Sect. 1.3).

(d) The locomotor movements made by the fish. Body and fin movements during swimming vary widely among fishes. The following classification of swimming modes is based on that of Breder (1926) as modified by Gray (1968) and Webb (1975a).

## 1.1.1 Use of Body and/or Caudal Fin

Breder (1926) distinguished three basic subdivisions within this general method of propulsion: anguilliform, carangiform, and ostraciform. In each type the main locomotor force is provided by lateral oscillations of the body and/or the tail. The oscillations are produced by alternate contraction and relaxation of muscles on either side of the body that result in backward pressure being exerted against the surrounding water. The three basic modes differ primarily in the proportion of total fish length that is involved in the oscillations. A fourth mode, the subcarangiform, is also recognized by some authors (e.g., Webb, 1975a). Figure 1.2 illustrates the four types.

*Anguilliform Mode.* The entire body and tail oscillate, with usually more than one complete wave-length within the length of the body. The amplitude of the oscillation is usually wide over the entire body length. The body is typically long, thin, and flexible. The best-known examples, from which the name for this mode is derived, are the eels (Anguillidae).

*Carangiform Mode.* The posterior portion of the body and the tail oscillate, with up to one-half a wave-length within the length of the body. The amplitude is largest at the posterior end. The body shape is fusiform, with a narrow caudal peduncle. This is a common mode of propulsion, found among many fast, efficient swimmers, such as herrings (Clupeidae), jacks (Carangidae), mackerels, and tunas (Scombridae).

*Ostraciform Mode.* The body is rigid and does not oscillate. Propulsion is by wig-wag oscillation of the caudal fin. Body shape is variable with little streamlining. Well-known examples are the boxfishes (Ostraciidae), puffers (Tetraodontidae) and the porcupine-fishes (Diodontidae).

*Subcarangiform Mode.* This mode is intermediate between the anguilliform and carangiform modes, and is distinguished as a separate category because it is found in many well-known groups of fishes (Webb, 1975a). The body and tail oscillate, with more than one-half a wave-length within the body length. The amplitude rapidly increases and is wide over the posterior one-half to one-third of the body. The body shape is fusiform. Typical examples are salmon and trout (Salmonidae) and cod (Gadidae). The intermediate position of this type emphasizes the fact that

3

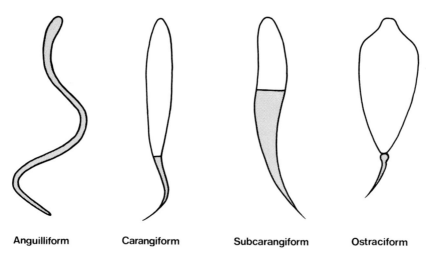

| Anguilliform | Carangiform | Subcarangiform | Ostraciform |

**Fig. 1.2.** Four basic swimming modes among fishes. The part of the body or tail that oscillates during locomotion is stippled. (After Webb, 1975a)

Breder's original classification was a somewhat arbitrary selection of three stages within a continuum ranging from extremely flexible eel-like forms to stiff, inflexible fishes. A common feature of the entire series is that caudal fin oscillations contribute a major portion of the propulsive thrust.

## 1.1.2 Use of Extended Median or Pectoral Fins

In these fishes the rays of the dorsal, anal, or pectoral fins oscillate in such a manner that waves of movement pass along the fins in a posterior direction. These waves are similar to the body and caudal fin movements in the anguilliform mode of swimming, but the wave amplitude is more constant along the entire length of the fin. Breder (1926) recognized four distinct types, based on the fin(s) providing the propulsive force (Fig. 1.3).

*Amiiform Mode.* The extended dorsal fin is the primary source of propulsion. Examples are the bowfin (Amiidae) and the African electric fish *Gymnarchus*.

*Gymnotiform Mode.* Similar to Amiiform, but with propulsion provided by the long, flexible anal fin. Examples are the knife-eels (Gymnotidae) and the electric eel *Electrophorus*.

*Balistiform Mode.* The soft dorsal and anal fins are used together to provide thrust. The best examples are the filefishes and triggerfishes (Balistidae).

*Rajiform Mode.* Most skates and rays (Rajiformes) live on or near the bottom and have large pectoral fins that undulate, especially at the outer edges, to produce

4

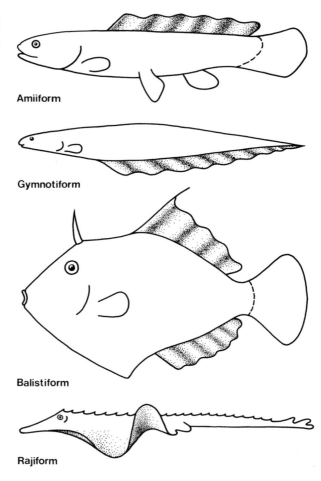

**Fig. 1.3.** Four swimming modes among fishes that use their median or pectoral fins for propulsion. (After Breder, 1926)

Amiiform

Gymnotiform

Balistiform

Rajiform

propulsive force. The eagle rays (Myliobatidae) and manta rays (Mobulidae) are pelagic swimmers and move by flapping their very large pectoral fins in unison, thus resembling the flapping flight of birds (Klausewitz, 1964).

## 1.1.3 Use of Shortened Median or Pectoral Fins

Three types of propulsion based on vigorous and rapid movements of relatively small pectoral or median fins were recognized by Breder (1926) (Fig. 1.4):

*Tetraodontiform Mode.* The small dorsal and anal fins are flapped from side to side, and the caudal fin is used for steering (Lindsey, 1978).

*Diodontiform Mode.* Propulsion is by the rhythmic undulation of the pectoral fins. Both this and the preceding mode are used by puffers (Tetraodontidae) and porcupinefishes (Diodontidae), although, as mentioned above, these fishes can also

5

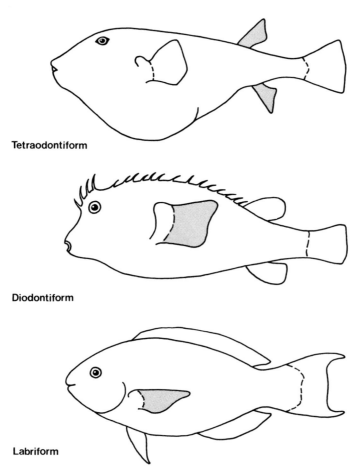

**Tetraodontiform**

**Diodontiform**

**Labriform**

**Fig. 1.4.** Three swimming modes among fishes that use small median or pectoral fins for propulsion. (After Nelson, 1976)

progress by the ostraciform mode of swimming, in which the wig-wagging tail provides the thrust.

*Labriform Mode.* The pectoral fins are moved in synchrony, acting like a pair of paddles. Some wrasses (Labridae) and parrotfishes (Scaridae) typically swim this way. They swing their pectorals forward flattened in the horizontal plane, then turn them through about 90° and pull them back, thus providing the forward propulsive force (Lindsey, 1978). Mudminnows (Umbridae) use a labriform pattern to advance slowly towards prey; the only apparent motion by the fish is the extremely rapid fluttering of the pectoral fins until the final leap at the prey is made in the carangiform manner.

In the above three propulsive modes the described fin movements provide slow or moderate forward propulsion. Most of these fish are also capable of short bursts of faster swimming by caudal fin oscillations, for example when frightened. Their

6

body and fin structures, however, do not in most cases permit sustained fast swimming. On the other hand, the sea perch *Cymatogaster aggregata* (and probably other embiotocids) can maintain continuous, moderate to high-speed swimming by using only the pectoral fins in the labriform mode. The fins appear to operate via a lift mechanism, and may usefully be compared with the wings of flying animals (Webb, 1973, 1975b).

## 1.2 Manoeuvering and Stabilizing

In addition to swimming forward, fish are capable of manoeuvering and making fine adjustments in orientation. They can thus maintain a controlled, yet dynamic stability in water. This capacity is especially important when feeding, avoiding predators, and interacting with conspecifics. Whereas the body and caudal fin provide the main propulsion for forward swimming in most species, the other fins are generally more important to all fish in controlling the finer stabilizing movements.

Teleost fins have two types of supporting structure; stiff, unbranched, and usually unsegmented *spines*, and flexible, branched, segmented *rays*. Adjacent units (spines or rays) are usually joined by a continuous membrane, although the spines of some species lack membranous support (e.g., sticklebacks, Gasterosteidae). Most fins can be spread open or folded, thus quickly changing the surface area exposed to water. However, the efficiency of fins in contributing to stability is based mainly on the flexibility and elasticity of the soft rays (Gosline, 1971). Flexibility is provided by the joints between adjacent segments of each ray, and elasticity is apparently contributed by the membranes associated with the internal structure of the rays. The spines are relatively inflexible and non-elastic, and their main roles are to provide support for the fins and defense against predation.

Efficient control of fine movements is achieved not only by adjustments of the fins but also by shifts and turns of the entire body. The three main types of body movement are conventionally named after the motions of a boat on the water surface (Fig. 1.5). These are: *yawing* movements, from side to side about the vertical body axis; *pitching* movements, about the horizontal, transverse axis; and *rolling* movements, about the horizontal, longitudinal axis. Control of these movements is largely automatic and is regulated primarily by impulses generated in the semi-circular canals (Gray, 1968).

Much of our knowledge of the ability of fishes to manoeuver is based on the work of Breder (1926) and Harris (1936, 1938, 1953). Reviews can be found in Marshall (1965), Alexander (1967), Gray (1968) and Lindsey (1978). Harris studied the behaviour of fish models in water and of fish with one or more fins amputated. He showed that *yawing* can be controlled by spreading the posteriorly placed, vertical median fins (dorsal, caudal, and anal), which thus act as "keels". Long-bodied fish achieve yawing stability with relatively small median fins, while shorter-bodied fish, and especially those with deep, highly compressed bodies, require larger median fins to maintain stability. Some fish use the flexible, soft-rayed

7

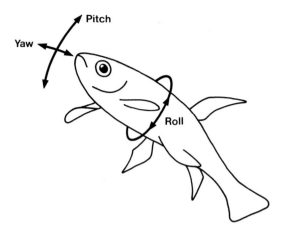

**Fig. 1.5.** The three main types of body movement in fishes. (Alexander, 1967)

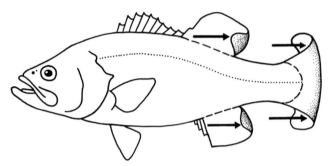

**Fig. 1.6.** Soft portions of median fins of a stationary largemouth bass, *Micropterus salmoides*, curved to opposite sides to act as a mid-water "anchor". (After Breder, 1926)

portions of the median fins to help maintain position in flowing water; the anal and dorsal fins may be curved to one side, the caudal to the other, the combination acting like a mid-water "anchor" (Fig. 1.6).

Control over *pitching* movements is more complicated. In general the paired fins are most useful. Pectoral fins placed well forward on the body can produce anterior lift or drop forces depending on their angle of inclination when held open (Fig. 1.7). In many negatively buoyant sharks the pectorals are used primarily for vertical movements; the posteriorly placed pelvics counter the pectoral fin forces, and return the body to the horizontal plane. In addition, many sharks have a heterocercal tail and large, strong pectorals shaped like hydroplanes. The shapes and locations of these fins enable sharks to control vertical movements, but also reduce their manoeuverability. In more modern fishes, whose swim bladders provide increased buoyancy, the paired fins and caudal are not needed primarily for lift, and can contribute to greater manoeuverability. This includes the abilities to start and stop quickly, turn sharply and back up, all of which sharks are poorly equipped to do.

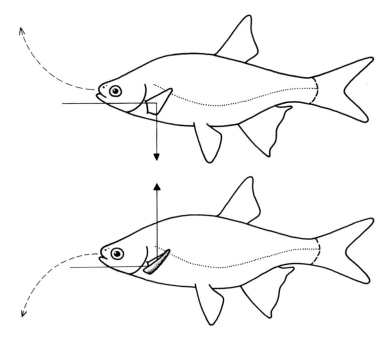

**Fig. 1.7.** Pectoral fins being used to control pitching of a swimming fish. *Solid arrows* indicate the vectors of force of water striking the fins. *Broken arrows* show the direction of pitch. (After Breder, 1926)

Quick starts are produced by caudal fin propulsion, but the paired fins are essential to control pitch during braking. Pelvics that are positioned posteriorly exert a downward as well as braking force when the fish stops. This tends to lower the tail and raise the head, forces that are opposite to those created by pectoral fins placed low on the anterior end of the body. However, the braking efficiency of the pectorals is greater if they are higher on the sides, and since in that position they create less downward force during braking, there is less need for posteriorly placed pelvics. Pectorals high on the body and pelvics placed well forward combine efficient braking with good pitch control in a "four-wheeled braking system which produces an extremely efficient and well-controlled stop" (Harris, 1953, p. 21).

Finally, side-to-side *rolling* movements are controlled mainly by the median fins, although the paired fins may be used to restore equilibrium to fish that have rolled far to one side. Such loss of roll control can happen, for example, to coral reef fishes while being "cleaned" by labrid cleaner-fishes (e.g., *Labroides*). During that behaviour the fish being cleaned often rolls over to one side; this is associated with immobility of the entire fish, including its fins (Eibl-Eibesfeldt, 1959).

In his pioneering work on fish locomotion Breder concluded that: "While the propelling and steering mechanism of fishes is intimately connected, there is a general tendency to place the burden of the first on the body muscles and that of the latter on the fins" (Breder, 1926, p. 286). This generality still holds, despite the recent increase in detailed fish locomotion studies, directed mainly at hydrodynamic and energetic aspects (e.g., Wu et al., 1975; Webb, 1975a).

9

## 1.3 Drag-Reducing Mechanisms

Drag is the resistance to motion of a body moving in a fluid (Webb, 1975a). Much attention has been paid to the mechanisms available to reduce drag and hence increase the swimming efficiency of fishes. These have been reviewed by Bone (1975) and Webb (1975a). Since most of them are concerned strictly with the hydrodynamics of individual fish propulsion, they are outside the scope of this book. However, two proposals that relate swimming in groups to drag reduction will be briefly considered.

### 1.3.1 Schooling Behaviour

Fish swimming in a polarized school [i.e., all of similar size, moving at the same speed and uniformly spaced (Shaw, 1970)] can reduce drag forces by using some of the energy in the wakes of fish ahead of them in the school. The effect is most clearly demonstrated in a flat, two-dimensional school, such as that often seen among small fishes in shallow water (Breder, 1959). Each individual sheds a double trail of vortices, rotating in opposite directions, so that directly behind it, the water movement (relative to the vortices) is in the opposite direction to the swimming fish. To either side, the induced water movements are in the same direction as the swimming fish (Fig. 1.8). Thus fish swimming in a diamond pattern, rather than with each one directly behind the closest preceding fish, can best utilize the energy in these vortex trails. The most efficient spacing is when the lateral distance between fish is twice the width of the vortex trail; each fish then benefits from the vortices of the two fish diagonally in front. Given optimal conditions, i.e., continual swimming in perfect synchrony of fish in an array of two-dimensional diamonds, savings of locomotor energy of up to a factor of five are theoretically possible (Weihs, 1973, 1975). Given this order of possible energy saving, together with data from real fish schools showing that the diamond spacing pattern is common in nature (Breder, 1965, 1976), it seems highly likely that some savings are actually realized, and hence this is a strong argument for the hydrodynamic advantage of fish schooling.

The locomotor advantages of swimming close to other fishes, and thereby reducing drag, may be responsible, at least in part, for the frequent association of small fishes with much larger individuals of other species (Breder, 1976). A well-known example is the pilot-fish, *Naucrates ductor*, that swims for long periods close to large sharks, manta rays, tunas, and even ships (Shuleikin, 1958; Marshall, 1965).

### 1.3.2 Mucous Production

The mucous secreted by cells in fish epidermis has important drag-reducing properties (Breder, 1976). The effect is not uniform across all species. Rosen and Cornford (1971) estimated the reduction in friction of a mixture of water and fish mucous, as compared to a control water solution, in a specially designed rheometer.

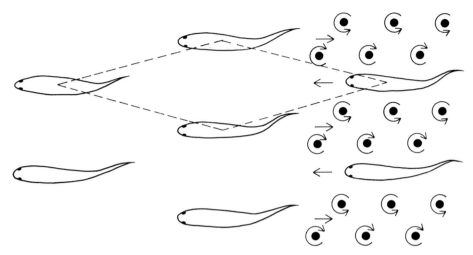

**Fig. 1.8.** Schooling fish from above in a hydrodynamically efficient diamond spacing pattern (*broken lines*). *Curved arrows* show direction of vortices set up by swimming fish; *straight arrows* the direction of induced water movement relative to the vortices. (After Weihs, 1973)

They found great interspecific variation. Mucous from the Pacific barracuda (*Sphyraena argentea*) was highly effective, producing a 65.9% reduction in drag in a 5% solution. Since the barracuda is a highly streamlined predator, capable of sustained high speeds while chasing prey (Hoyt, 1975), the friction-reducing property of its mucous seems to be an adaptation contributing to its predatory hunting mode.

Breder (1976) suggests that mucous production has locomotor benefits not only for the fish that secretes it, but also for those swimming close behind, because the vortex trails behind each fish contain small amounts of mucous from that individual. He suggests there may be a lubricity gradient in the water through which a school is passing, being lowest at the front, where the leading fish have only their own mucous to reduce drag, and highest at some point behind them, depending on school size. Thus leading fish may fatigue faster than the others and soon be replaced by them. This may help to explain the frequent "churning" seen in schools, as fish at the leading edge are replaced by those from behind (Breder, 1976).

## 1.4 Locomotion in Air

Although most fishes spend their lives completely submerged in water, some occasionally travel briefly through the air. These include fish that jump out of water in pursuit of aerial prey, fish that jump through the air while surmounting obstacles during migration, fish that skitter along the water surface while fleeing from predators, and finally the so-called "flying fishes", that regularly travel many metres out of the water. Here I am concerned only with the latter two categories,

11

since these are the fish that have both structural and behavioural modifications promoting aerial locomotion.

A fish in air is immediately subject to a greatly increased gravitational force, because it no longer has the buoyancy provided by water. The main structural adaptations to counter this increased gravitational pull are the enlarged paired fins. The best-known flying fishes are tropical and subtropical oceanic Exocoetidae. They have either the pectoral fins, or both the pectorals and pelvics, greatly enlarged. These fins, when fully opened, provide the fish with lift after leaving the water (Hubbs, 1933, 1937). The fish gain initial momentum by swimming rapidly as they approach the water surface. Once airborne, fishes with only their pectoral fins enlarged, open them and begin to glide. An example is *Exocoetus volitans* (Fig. 1.9). The direction of movement in the air is not well controlled, and these fish are easily blown about by cross-winds (Hubbs, 1933).

Aerial locomotion is better controlled in species that use both sets of paired fins for lift. The best-known are fishes of the genus *Cypselurus* (Fig. 1.9). On leaving the water the large pectoral fins immediately expand fully, but the tail continues to oscillate in the water as the fish taxis along the surface, slightly head-up and with gradually increasing speed. The extended lower caudal lobe enhances the taxiing form of surface locomotion. When the tail is lifted completely from the water the fish assumes a more horizontal orientation and the pelvic fins open, to provide four-fin lift to the gliding fish. The successive stages in such a flight are shown in Figure 1.10. These fish have considerable control over the direction of their "flights", both in the vertical and horizontal planes. They have often been seen turning to avoid contact with a ship, and rising and falling to remain close to the water surface when waves are high (Hubbs, 1933).

There has been some controversy over the role of the pectoral fins in propulsion after a flying fish leaves the water. Some naturalists claim to have seen the tips of the pectorals vibrating, and assumed they contribute to forward propulsion (Hubbs, 1933). However, after careful observation of many flights of exocoetids, Breder (1926), Hubbs (1933), and Myers (1950) have all concluded that fin tip vibrations cease when the fish becomes fully airborne. Slight pectoral movements, if present, are caused by vigorous tail oscillations while the fish taxis at the surfaces (Fig. 1.10C). Equally important, the cypselurines lack the massive musculature and strong pectoral skeletal features that would be required if the pectoral fins were actively contributing to propulsion in the air (Gray, 1968). Thus it seems clear that the movements of exocoetid fishes through the air are in the form of extended glides, not true flapping flight.

The role of the caudal fin in aerial locomotion is clearly seen in some open ocean species as the airborne fish descends after its initial glide. The fish may fold its pelvic fins and touch water first with the ventral caudal lobe. The tail then begins rapid oscillations, the fish gains speed and takes off again in another gliding flight. Although a single glide, followed by re-entry of the water, is most common, occasionally up to five take-offs, separated by brief caudal contacts with the water, have been seen (Marshall, 1965).

Among freshwater fishes that are said to fly in air there is some evidence, although mostly indirect, that they perform flapping flight. The South American hatchetfishes (Gasteropelecidae) often skitter along the water surface when

12

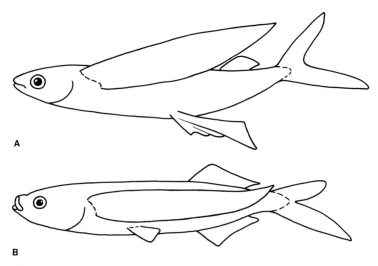

**Fig. 1.9.** **A** *Cypselurus lineatus*, a flying fish with enlarged pectoral and pelvic fins. (After Marshall, 1965). **B** *Exocoetus volitans*, with enlarged pectoral fins. (After Marshall, 1964)

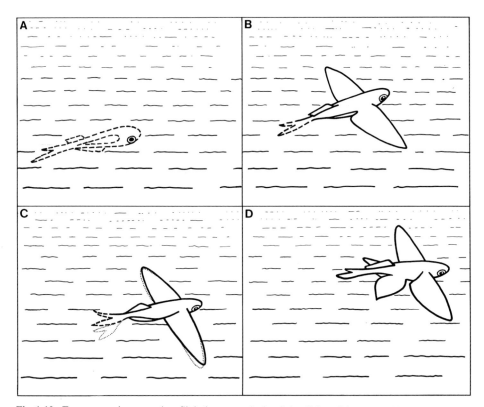

**Fig. 1.10.** Four successive stages in a flight by a cypselurine flying fish. **A** fish approaching water surface with paired fins folded, **B** pectoral fins spread as fish breaks through the surface, **C** taxiing at the surface, with caudal fin in the water, **D** pelvic and pectoral fins spread as fish becomes fully airborne. (After Hubbs, 1933)

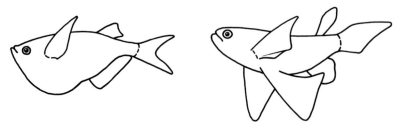

**Fig. 1.11.** Two freshwater "flying fishes". *Left, Carnegiella* sp.; *right, Pantodon buchholzi*. (After Marshall, 1965)

disturbed, with the body and tail in contact with the water. The large pectoral fins beat rapidly against the water surface, thus presumably contributing to forward propulsion. Such aerial excursions may continue for up to 15 m, with the fish then becoming completely airborne for a distance of 2 to 3 m before falling back into the water (Eigenmann, 1912). One member of this family, *Carnegiella*, makes a buzzing sound, probably with its rapidly beating pectorals, when airborne (Weitzman, 1954). These fishes have a highly compressed body with a deep pectoral skeletal area and large associated pectoral fin muscles (Fig. 1.11), and it seems highly probable that the fins do provide some propulsive force to the fish while in the air.

The highly specialized West African freshwater butterflyfish (Pantodontidae) may also be capable of true aerial flight, but again the evidence is indirect. *Pantodon buchholzi* has highly developed pectoral musculature and girdle (Fig. 1.11). The fins cannot be folded back against the body, but can be moved in the vertical plane, with especially strong downward strokes (Greenwood and Thomson, 1960). Like some gasteropelecids, *P. buchholzi* has been described as beating its pectorals against the water while skimming along the surface.

Conclusive evidence on the means of aerial propulsion is not available for any of these freshwater "flying fishes". Since they move so quickly, and stay so close to the surface after leaving the water, high-speed photography is probably necessary to establish their mode of progression in the air. It seems clear that they do not simply glide, as the oceanic exocoetids do; nor do they have control over their flight direction after becoming fully airborne.

Locomotion by fishes in the air seems to be associated with escape from approaching predators, ships, or other disturbance. Many fishes are capable of undirected leaps out of the water, as one means of avoiding predator attacks. The fish discussed in this section are also able to move horizontally in the air, and this capability reaches its peak in the prolonged, gliding flights of the oceanic exocoetids.

## 1.5 Locomotion on Land

A number of species of fish can move about on land, often some distance away from water. Most use much the same mode of locomotion as they do when swimming. For example, juvenile eels (Anguillidae) on their upstream migrations in fresh

14

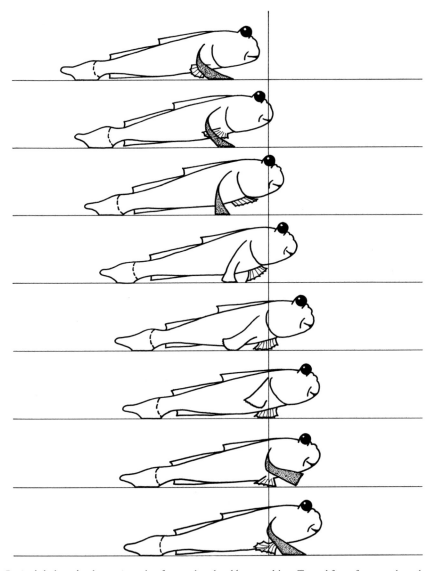

**Fig. 1.12.** *Periophthalmus koelreuteri* moving forward on land by crutching. Traced from frames selected from a film to show progressive changes in positions of pectoral and pelvic fins. (Modified after Harris, 1960)

water, and adult eels when moving downstream to the sea, can by-pass waterfalls and dams by leaving the water and moving overland. Propulsion is by anguilliform undulations of body and tail, although the bending of the body is usually exaggerated because of the lower density of air than water. If the fish can brace itself against a rock or other object on the substrate the undulations are more like those in the water (Gray, 1968). Other species, such as the catfishes (*Clarias* and *Saccobranchus*) and the snake-heads (*Ophicephalus*) progress on land by exaggerated carangiform-type lateral undulations of body and tail. These fishes

15

move onto land mainly to escape from deteriorating water conditions, in search of better quality water.

Only a few species are capable of a truly amphibious existence, staying out of water for prolonged periods. The best-known are the mudskippers (*Periophthalmus*). These fish dig burrows in the soft, muddy substrate of mangrove swamps and mud flats of tropical Africa, Asia and Australia. They use two different means of propulsion on land (Harris, 1960; Stebbins and Kalk, 1961):

*Ambipedal Progression.* This is a slow form of locomotion, somewhat analogous to a human with crutches, hence also called "crutching" (Fig. 1.12). The strong pectoral fins are folded to form sturdy struts, and they act together to support the body off the ground as it moves forward. Then the short, stiff pelvic fins hold the body off the substrate while the pectorals are lifted, swung forward and placed on the ground again. Thus the pectoral fins act like crutches and the pelvic fins like legs. The fish cannot lift its tail, which drags along the ground as the fish progresses. Mudskippers can move over mud flats and onto the exposed prop roots of mangrove trees by using the crutching mechanism.

*Skipping.* This is a much faster form of locomotion than crutching. The propulsive force is provided by the tail, which is bent to one side and then suddenly straightened with the stiff ventral caudal rays pressing into the mud. The pelvic fins, which are closed at the same moment, raise the head off the ground. The pelvics thus provide the vertical component, and the tail provides the forward component to the skip, during which the fish jumps clear of the substrate. Skipping is usually performed when the mudskipper is alarmed, and moving to its burrow or into water at maximum speed.

Some *Periophthalmus* species can climb up the branches of mangroves. For example, *P. chrysospilos* holds onto branches by using its pelvic fins, that are fused into a single suctorial disk, and also by grasping the branch on either side with its strong, flexible pectoral fins (MacNae, 1968).

In conclusion, on the grounds of locomotor efficiency alone, one might expect natural selection to favour trends in fish morphology that contribute to: (1) reduction in energy output for swimming; (2) increased capacity for acceleration and speed; and (3) increased manoeuverability and stability (Alexander, 1967). In fact, these trends must have interacted with other demands, e.g., for camouflage, for access to hiding places, for long distance migrations, and for efficient communication during intraspecific social interactions. The incredible array of shapes and sizes among fishes reflects the interplay among these various selection pressures, taking place in a highly buoyant medium. Locomotion is an important capacity, but is not an end in itself. Its most general contribution is that it enables fish to meet their other primary requirements, that is, to acquire food, avoid enemies, and breed.

Chapter 2

# Feeding Behaviour

The acquisition of food by fishes is a process that usually involves: searching, detection, capture, and ingestion. A hungry fish performs searching activities that increase the probability that it will discover food. Initiation of searching depends primarily on the "hunger state" of the animal, which in turn is believed to be controlled by interaction between the amount of food in the stomach and the systemic need (metabolic debt) of the fish (Colgan, 1973). On detecting potential food items, the fish orients towards them, approaches, and attempts to capture and ingest them, or parts of them. For herbivores the capture may consist of nothing more complex than biting pieces out of plants, nibbling or scraping algae from a hard substrate, or straining phytoplankton from water as it passes into the mouth and out past the gills. Food capture by carnivores generally requires more elaborate techniques because potential prey organisms have a wide range of behavioural and structural adaptations for avoiding capture.

Further complexity in fish feeding patterns stems from the fact that many species are omnivorous. Some are clearly opportunistic, varying their diet quickly to capitalize on sudden, short-term abundance of particular prey. Some, including many coral reef fishes usually considered to be food specialists have, on closer examination, been found to incorporate into their diets items from several trophic levels (Fishelson, 1977). Over their lifetime, many fishes change the main components of their diet as they themselves increase in size and are able to handle different forms of prey.

Despite these sources of variety in diet, it is convenient to review the feeding behaviour patterns of fishes by describing the major trophic categories. Emphasis is on description of the common foraging strategies, the sensory modalities used in feeding, and on the relationships between feeding behaviour and habitat. Although the subject cannot be discussed without some reference to structural adaptations for feeding, these are not considered in detail here. Several reviews of fish morphology in relation to feeding are available (e.g., Keast and Webb, 1966; Alexander, 1967; Gosline, 1971; Fryer and Iles, 1972).

## 2.1 Detritivores

Organic detritus has been defined as "...particulate organic material originating from dead bodies of organisms or from non-living fragments and excretions shed by living organisms..." (Odum and de la Cruz, 1963, p. 39). It accumulates on the substrate of any aquatic environment where water movement is reduced. Detritus deposits probably always contain silt, sand or other non-organic particles that also collect in such areas, as well as bacteria, algae and other unicellular organisms. A wide variety of invertebrate organisms may also occur in detritus.

Stomach analyses have shown that many species of fish ingest detritus, but if a stomach also contains whole or recently fragmented invertebrates, or quantities of algae, it may be impossible to determine whether the detritus was swallowed incidentally or for its own nutritive value. Fish whose stomachs contain little else but detritus (and the inevitably associated bacteria and microfauna) are assumed to be feeding directly on it. Examples from freshwaters are several *Tilapia* species in the African Great Lakes and the Grand Lac of Cambodia (Hickling, 1961), some cyprinids, such as *Cyprinus* spp. and the barb *Puntius bimaculatus* (De Silva et al., 1977), and a number of catostomids (Hynes, 1970). Marine detritivores include some coral reef blennies and gobies (Hiatt and Strasburg, 1960), the milkfish (*Chanos chanos*) and some mullets (*Mugil* spp.) (Odum, 1970).

One can hardly speak of a foraging strategy for detritivorous fishes. As far as is known, they swim along near the bottom, sucking in loose surface material, spitting out larger particles and swallowing the rest. *Mugil cephalus* is known to use its pharyngeal filtering device, consisting of gill rakers and teeth, to sort out and expel the larger fragments (Odum, 1970). Tactile and possibly also chemical receptors in the mouth and pharynx probably function in this sorting process.

The nutritional value of ingested detritus appears to come from both the decomposing organic particles and the bacteria and other micro-organisms attached to the particles. Some African fishes feed actively on partly digested water lily leaves excreted by herbivorous fishes (Hickling, 1961), and others (*Tilapia* sp.) have the remarkable habit of eating detrital deposits composed largely of decomposing hippopotamus faeces (Fryer and Iles, 1972). In the latter case, crude protein levels in the rectum contents of the fish were 60% less than in their stomach contents, showing that the ingested materials were being used as food (Fish, 1955). The bacteria and protozoa on the detritus particles eaten by *Mugil cephalus* in salt marshes became much less abundant as the material passed through the digestive tract, suggesting that they were a source of nutrients for the fish (Odum, 1970).

Finally, a little-known, but probably common food source for many coral reef fishes is the mucous produced by coral polyps, especially in response to disturbance or injury. Although some fishes feed directly on polyps, others (including some chaetodontids, pomacentrids, and monacanthids) have been seen picking at the mucous where it collects in the clefts among the protruding skeletal parts of a coral colony. When pieces of coral were broken off, or the surface of a colony rubbed vigorously, many small fishes quickly gathered and fed on the copious clouds of flocculent mucous that were produced in response (Benson and Muscatine, 1974). Coral mucous is rich in certain lipids, and is likely an important energy source for fishes that feed regularly on it (Benson and Lee, 1975).

Thus, detritus in various forms can be a valuable source of nutrients for fishes with appropriate sorting mechanisms in the buccopharyngeal area, and digestive systems that can utilize the variety of organic matter occurring in it.

## 2.2 Scavengers

Some fish are scavengers, feeding on dead and dying organisms, including other fishes caught in nets, traps, or on hooks. Discovery of such food is probably mediated by chemical and perhaps auditory cues (in the case of struggling captive animals), since dead organisms are usually on the bottom, often at depths where visual cues are faint.

Sharks have a reputation for indiscriminate scavenging because a wide variety of refuse, including bottles, cans, papers, clothing, and furniture, has been found in shark stomachs (Lineaweaver and Backus, 1970; Budker, 1971). This reputation is probably much exaggerated, because many sharks whose stomach contents have been examined were captured near harbours or in narrow shipping lanes, where refuse is more common than in the open sea. On the other hand, some smaller sharks, such as the dogfish (*Squalus acanthias*) are well-known for eating fish caught in commercial fishing gear (Lineaweaver and Backus, 1970; Hart, 1973). Most sharks should probably be considered opportunistic feeders, scavenging when the opportunity arises (Springer, 1967).

Many smaller fishes also scavenge opportunistically, at least when in captivity. Tropical fish hobbyists are familiar with a dead or dying fish being nibbled and eventually consumed by other fishes in an aquarium. The extent of such feeding in nature is unknown; once a dying fish drops to the bottom it is usually soon consumed by benthic invertebrate scavengers.

However, one group of fishes scavenges regularly. These are the hagfishes (Myxinidae). They are benthic marine animals with rudimentary eyes that feed on dead or trapped fish, and also on worms and gastropods (Denison, 1961; Strahan, 1963a,b). Food is detected by chemical cues, to which the hagfish respond with positive rheotaxis, swimming upstream against currents until the source of chemical stimulation is found (Strahan, 1963a). They bite at prey with their numerous lingual teeth and single median palatal tooth (Dawson, 1963). Once the skin is penetrated the hagfish burrows into its prey, eating as it goes; hence they are sometimes found inside fish captured on the bottom. A hagfish can quickly form a knot with its own tail and body and, by pressing the knot forward, can exert considerable pressure against a substrate, while retracting its head (Fig. 2.1). This allows it to pull with some force on food grasped by the teeth, and to pull itself backward out of the body of a dead fish on which it has been feeding (Strahan, 1963a).

## 2.3 Herbivores

Three general types of herbivorous fishes can be recognized. These are: *grazers*, that crop algae so close to the underlying substrate that some of the inorganic materials may be taken into the mouth with the algae; *browsers*, that bite off pieces of plant above the substrate; and *phytoplanktivores*.

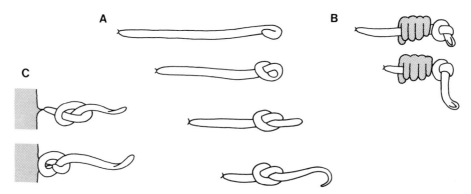

**Fig. 2.1.** Behaviour of the hagfish *Myxine*. **A** self-knotting, **B** escape from human grasp by knotting and pulling backwards, **C** pulling on food by biting and knotting. (After Strahan, 1963b)

## 2.3.1 Grazers

These fish remove algae from the substrate by scraping or rasping at the surface and sucking the loosened materials into the mouth, or by biting pieces out of the substrate. Well-known marine grazers are the parrotfishes (Scaridae) of coral reef communities. They remove algae either by scraping or biting at the reef surface with their hard, beak-like mouth, and then grinding the loosened material in a specialized pharyngeal mill (Randall, 1968). Quantities of coral rock, shell fragments and sand may be ground up and swallowed along with the organic material, and then passed out with the faeces. Parrotfishes are thus responsible for producing large amounts of fine reef sediments (Stephenson and Searles, 1960; Bardach, 1961). Fragments of coral polyps may also be eaten in this process, but if the proportion of algae to polyps in the stomach contents is consistently high, the fish are classed as grazing herbivores (Hiatt and Strasburg, 1960).

A school of actively grazing parrotfishes produces clearly audible sounds. Sartori and Bright (1973) found that juveniles and adults less than 255 mm long fed by rasping at the reef surface with the teeth in a rapid, rhythmically repeated manner. Larger fish took irregularly spaced and more deliberate bites from the reef. These two feeding modes produced distinctive sounds labeled Scrape and Crunch respectively. Automatic recording of these sounds could be used to monitor feeding activity at selected locations and thus estimate the amount of calcareus material removed from the reef by the grazing fish (Sartori and Bright, 1973).

Many freshwater fishes have specialized mouth parts that enable them to remove firmly attached algae from the substrate. For example, some Latin American catfishes (genus *Plecostomus*) have a sucker-like arrangement of the skin around the ventrally placed mouth, by which they can securely anchor themselves to rocks. Their long, flexible, bristle-like teeth are then used to rasp algae from the rock surface (Alexander, 1967). These fish are widely used by hobbyists as aquarium "algae-eaters". The North American cyprinid, *Campostoma anomalum*, is called the "stoneroller minnow" because it turns stones over and with the cartilagenous ridges surrounding the ventral mouth, scrapes encrusting algae off the stones (Hynes, 1970).

20

Some of the most detailed studies of fish grazing have concerned the so-called "Aufwuchs-eaters". The term "Aufwuchs" has long been used by German limnologists (and is now common in the English-language literature) for the mat of algae and associated organisms that forms on rock surfaces in freshwater (Hickling, 1961). A great variety of specializations in mouth and teeth of African lake cichlids that feed on Aufwuchs is described by Fryer and Iles (1972). Teeth of different shape, size, and numbers are used to scrape, file, comb, or nibble algae from the rocks. The remarkably mobile mouth and lips of some species contribute to efficient algae-grazing. For example, in *Pseudotropheus zebra*, a rocky shore inhabitant of Lake Malawi, "The mobility of the mouth is such that when it is opened and the tooth-covered 'lips' are pressed against a rock surface they mould themselves to the shape of the rock and, as the mouth closes, the teeth comb algae from its surface" (Fryer and Iles, 1972, p. 68).

Some fishes also graze on algae growing epiphytically on plant leaves. For example, *Hemitilapia oxyrhynchus* of Lake Malawi has dentition that allows it to grasp a *Vallisneria* leaf in its mouth and remove the algae by nibbling as it moves slowly along the leaf (Fryer and Iles, 1972).

## 2.3.2 Browsers

There is little in the way of behavioural specialization for browsing among fishes. Species that feed regularly on leaves and stems of higher plants, or on algal fronds have a variety of mouth and teeth adaptations for efficiently biting pieces out of plants. The best-known examples are from tropical waters, both freshwater and marine.

The grass carp (*Ctenopharyngodon idella*) bites the leaves off grasses growing in ponds so efficiently that it has been proposed as a biological control agent for areas where aquatic weed control is desired (Cross, 1969). It has no teeth in the mouth but the pharyngeal apparatus is such that the fish can grasp a leaf and tear pieces from it by vigorously shaking its head. Some *Tilapia* species, referred to as "leaf choppers", have several rows of bi- and tricuspid teeth, with the outer rows having a sharp cutting edge (Fryer and Iles, 1972). The South American characid *Myleus* uses the sharp edges of the teeth in its upper and lower jaws to bite pieces out of submerged leaves as though cutting with scissors (Alexander, 1967). Many tropical freshwater fishes live in streams subject to extreme seasonal flooding, which inundates wide areas rich in plant life. These plants begin to decay underwater and are then eaten by a variety of herbivores (Hickling, 1961). Some coral reef species also have teeth adapted for biting off the fronds of algae that grow profusely in association with corals. These include some surgeon fishes (Acanthuridae), damselfishes (Pomacentridae), triggerfishes (Balistidae) and rabbitfishes (Siganidae) (Hiatt and Strasburg, 1960; Jones, 1968).

## 2.3.3 Phytoplanktivores

A number of fishes feed primarily on phytoplankton, including diatoms, dinoflagellates, green and blue-green algae. Others only occasionally use it as food, for example, when temporary blooms occur. Filter-feeding is the primary method of collection. The fish swims with open mouth, straining organisms from the water as it flows past the gills. The filtering is done by gill rakers, and some species (e.g., *Tilapia*) also have fine projections called microbranchiospines on the gill arches. These carry even finer spines (36 to 40 $\mu$ long) which together form an extremely fine-meshed filtering system, capable of retaining the smallest phytoplankters and even large bacteria (Hickling, 1961). However, since microbranchiospines occur also in species that feed mainly on higher plants and zooplankton, their main function may be to protect the delicate gill filaments from abrasion by sand, silt, and other particles taken in with food (Fryer and Iles, 1972). Some phytoplanktivores also produce mucous in the mouth that helps to trap fine food particles which are then swallowed as a bolus (Fryer and Iles, 1972).

Several of the largest fishes in the world, including whale sharks (Rhincodontidae), basking sharks (Cetorhininae) and manta rays (Mobulidae), are planktivores. These huge animals consume very large numbers of organisms, but there is no evidence that they distinguish between phyto- and zooplankton as they feed.

Summing up, the primary feeding adaptations of herbivorous fishes are structural, not behavioural. Since plants have no escape mechanisms, fish require no specific capture strategies to utilize them as food. Provided their mouth, teeth, gill rakers, and digestive tract are appropriate for cropping, swallowing, and digesting plant material, they will use plants as sources of nutrition.


## 2.4 Carnivores


Most fish are carnivorous, that is, they eat live animal prey, at least during some stage of their lives. The great diversity in aquatic animals that can be utilized as food is reflected in a wide variety of carnivorous feeding strategies. Many prey organisms, especially aquatic insects and zooplankton, have short life cycles with brief periods of maximum vulnerability. Hence, fishes feeding on them must frequently change their diet. Also, as fish grow, their capacity to capture, handle and ingest mobile prey increase, allowing for still further variety in diet. Despite this potential for dietary breadth it is convenient to discuss carnivorous feeding methods according to major categories of prey. I consider below four such categories: benthos, zooplankton, aerial prey, and fish.


### 2.4.1 Benthivores

This section deals with feeding strategies aimed at benthic organisms other than fish. Benthic invertebrates are adapted to avoiding predation by firm attachment to the substrate, quick withdrawal into a protective covering, camouflage, presence of spines, nematocysts or toxic compounds, or by combinations of these. Their most

22

general characteristic is a high degree of localization; many are sessile, most others move only short distances from their usual resting places. Thus foraging benthivores are specialized for detecting and handling prey in various ways, rather than for chase and capture.

### 2.4.1.1 Picking at Relatively Small Prey

Many diurnally active fishes feed by picking one or a few small organisms at a time from the substrate. Their searching behaviour clearly shows the importance of vision in this process. They scan the bottom while holding position or swimming slowly, then fixate visually on one locality, quickly dart and snap at that spot, and resume searching. Sometimes the prey is spit out and snapped up one or more times, as the fish separates the food from inedible particles, or breaks it into smaller pieces for easier swallowing. This scan-and-pick method is used by some stream-living salmonids, for example, juvenile Atlantic salmon (*Salmo salar*), that feed on immature, aquatic stages of insects (Kalleberg, 1958; Keenleyside, 1962).

Vision also mediates feeding responses in several species of darter (Percidae), small stream-living fishes that feed by picking invertebrates from the substrate (Turner, 1921). Two experimental studies, involving seven species, showed clearly that visual but not olfactory cues from tubificid worms stimulated feeding snaps by darters (Roberts and Winn, 1962; Daugherty et al., 1976).

Many benthivorous fishes have highly modified mouth and teeth that enable them to pick small organisms from the substrate with efficiency. For example, the small African cichlid *Labidochromis vellicans* has a narrow mouth with very long, curved teeth at the tip of both jaws that act as fine forceps, well-designed for picking up small prey (Fryer and Iles, 1972). Several other African cichlids have large fleshy lobes on the mouth that are presumably sensitive to tactile stimulation by prey. *Haplochromis euchilus* feeds by moving slowly with its large lips pressed against a rock, then picking up prey with its teeth (Fryer and Iles, 1972). The marine carangid *Gnathanodon speciosus* picks small invertebrates out of algal mats by slowly working its protrusible, toothless jaws through the algae until it encounters suitable prey (Hobson, 1968a). Many fishes living on coral reefs feed by picking individual prey organisms off the coral or from algae growing on the coral (Hiatt and Strasburg, 1960). Extreme specialization for this feeding mode is shown by some of the chaetodontids (butterflyfish) that have long, narrow protruding snouts with which they snap up small organisms hiding in crevices among the coral (Eibl-Eibesfeldt, 1975). Observation of these fish while feeding strongly suggests that they visually fixate on individual prey before snapping. Some chaetodontids use their extended snouts and fine, protruding teeth to bite off the tips of individual coral polyps (Hiatt and Strasburg, 1960; Reese, 1977). Although coral colonies as a whole are sessile, the polyp feeders must approach slowly and bite quickly in order to seize the polyp tip before it is withdrawn into its protective skeletal mass.

Bottom-living flatfishes (Pleuronectiformes) are also capable of snapping up individual prey from the substrate. Their foraging strategy may include a preliminary stalking phase. For example, the summer flounder (*Paralichthys dentatus*) captures small shrimp by visual fixation, stalking by slow "crawling" movements along the bottom, using the dorsal and anal fins for propulsion, then

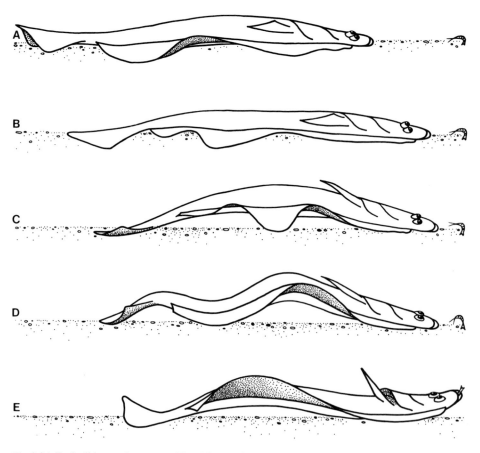

**Fig. 2.2A-E.** Stalking and capture of benthic prey by *Paralichthys dentatus*. (After Olla et al., 1972)

arching the body, bearing downwards vigorously with the caudal fin, and leaping at the prey with open mouth and engulfing it (Fig. 2.2; Olla et al., 1972).

Closely related species may use somewhat different strategies for the same food resource. A comparative study of feeding behaviour in adult cutthroat trout (*Salmo clarki*) and Dolly Varden char (*Salvelinus malma*) from the same lake showed that the latter were more efficient at capturing benthic chironomid larvae (*Psectrotanypus* sp.; Schutz and Northcote, 1972). Visual scanning of the bottom was the basic search technique of both species, but at the highest light intensities used, Dolly Varden swam faster, closer to the bottom, and more continually; cutthroat trout swam more slowly, further from the bottom and alternated between swimming and hovering more often. In addition the char often swam with pectoral fins brushing the bottom, suggesting the use of tactile receptors to detect partially buried prey. At lower light intensities both species searched close to the bottom, but cutthroat trout stopped searching and moved up into midwater sooner than Dolly Varden. Other observations showed that the trout were more efficient than the char at capturing surface prey. Thus, specialization for feeding on benthic prey by Dolly

24

Varden and surface prey by cutthroat trout appears to be an important aspect of the co-existence of the two species in the same small lake (Schutz and Northcote, 1972). Such divergence in feeding strategies may be common among closely related sympatric species.

### 2.4.1.2 Disturbing, then Picking at Prey

Some fishes uncover organisms buried in soft substrate by directing jets of water at the bottom with the mouth or fins and then snapping up exposed prey. The brown triggerfish (*Sufflamen verres*) uses this technique when feeding over sand, but when foraging on rocks the same species often tilts over on one side and rapidly undulates its dorsal and anal fins, thus causing a current that clears sediments from the rock surface and exposes small invertebrates (Hobson, 1968a).

Another strategy is for fish to watch others digging in the substrate and then dart in to feed on the exposed prey. Examples are small freshwater cyprinids which can be seen following catostomids that are sucking up and sorting out bottom materials. The same technique is used by many coral reef species. Goatfishes (Mullidae) typically feed by grubbing for prey in soft sediments with their mouth and barbels. A school of foraging goatfish is often joined by a variety of other species, including labrids, siganids, gerreids and acanthurids, all feeding on prey organisms disturbed by the digging goatfish (Hobson, 1968a, 1974; Fishelson, 1977).

The stirring-up of loose substrate materials can also provide food for conspecifics. Adult convict cichlids (*Cichlasoma nigrofasciatum*) guarding their free-swimming offspring often settle onto the bottom and vigorously undulate body and pectoral fins, thus stirring up a cloud of fine sediments. The young fish quickly gather and feed on particulate matter, including small invertebrates, disturbed by the parental digging (Williams, 1972).

### 2.4.1.3 Picking up Substrate and Sorting Prey

A number of benthivorous species use the goatfish technique, mentioned above, to capture invertebrates hidden in soft substrate. For example, the freshwater suckers (Catostomidae), have large, protrusible ventral mouths, with which they suck up quantities of bottom materials. Little is known about the sorting mechanism inside the mouth; possibly both tactile and chemical cues are used to separate edible from inedible material. The New World cichlid genus *Geophagus* (i.e., earth-eater) was so-named because of its habit of thrusting the strong, extended snout deeply into the substrate, withdrawing it, and vigorously chewing the mouth contents before ejecting inedible particles and swallowing the rest. The bottom of an aquarium containing several *Geophagus* soon becomes covered with pits created by the fish foraging in this way.

The Lake Malawi cichlid *Lethrinops furcifer* feeds mainly on chironomid larvae buried in the sand. It captures them by plunging its mouth into the substrate, filling it with sand, withdrawing and ejecting the sand, mostly through the opercular openings (Fig. 2.3A,B,C). Gill rakers screen these openings and retain the larvae as well as larger sand grains, which are often found in the gut with the chironomids. Another species of *Lethrinops* feeds by the same method, but retains somewhat

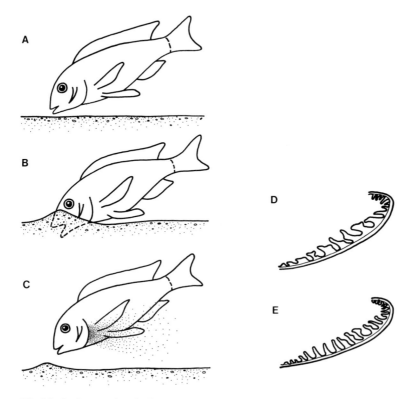

**Fig. 2.3.** *Lethrinops furcifer* feeding on buried prey (**A, B** and **C**). Outermost row of gill rakers of *L. furcifer* (**D**) and of *Lethrinops* sp. (**E**). (After Fryer, 1959)

smaller prey, in particular small ostracods, apparently because of its finer and more numerous gill rakers (Fig. 2.3D,E). This technique leaves series of pits and mounds on the bottom. "Literally thousands of such disturbances can be seen after a period of calm weather on sandy beaches where *L.furcifer* and its relatives are to be found, and indicate clearly the thorough way in which the habitat is searched for food" (Fryer and Iles, p. 79).

### 2.4.1.4 Grasping Relatively Large Prey

Many species fall into this category, but for the majority no detailed information is available on the foraging strategy used. An exception is the extensive work of Hans Fricke on the feeding behaviour of reef fishes. Among the most interesting of his discoveries is the range of techniques used to feed on sea urchins (genus *Diadema*). Fish of at least three families (Balistidae, Labridae, Lethrinidae) are capable of avoiding injury from the many long, sharp spines while feeding on *Diadema* (Fricke, 1971, 1973a). The triggerfish *Balistes fuscus* blows water at the urchin until it is turned over, then kills it by biting at the vulnerable oral disc. *B. fuscus* will move a variety of objects (including natural and man-made materials) in order to uncover a partly hidden urchin (Fricke, 1975a). The smaller species *Balistapus undulatus* bites off the urchin's spines, then lifts it by the stumps, drops it and quickly bites

26

at the oral disc while it is falling through the water. Two labrid species kill an urchin by pushing it over, picking it up by the body case, and vigorously beating it against a rock until it breaks into pieces. Finally a snapper (*Lethrinus* sp.) and another labrid simply attack an urchin directly by biting at it and eating the entire animal, including the spines (Fricke, 1973a).

Molluscs are handled in a variety of ways by carnivorous fishes. Some triggerfish (*Balistes*) use their strong, chisel-like teeth to bore through the shell and extract the meat of stationary forms such as oysters and mussels (Norman and Greenwood, 1975). The African cichlid *Haplochromis sauvagei* feeds on gastropods by grasping the snail's foot in its jaws and levering out the body, with the shell used as a fulcrum (Fryer and Iles, 1972). The small shark *Scyliorhinus canicula* can extract the whelk *Buccinum* from its shell by turning the animal over, grasping the foot with its strong, sharp teeth and shaking vigorously until the whelk comes loose from its shell (Brightwell, 1953). The mussel *Mytilus edulis* is the main food of the tautog, *Tautoga onitis*, who grasps one or more mussels with its anterior teeth, tears them loose from the substrate by vigorous head-shaking, then crushes them in its pharyngeal mill before swallowing them (Olla et al., 1974).

Most of the above examples of benthic foraging have concerned diurnally feeding fishes. However, many species feed at night, when much of the aquatic invertebrate fauna is most active (Starck and Davis, 1966: Hobson, 1968b, 1972, 1974). Little is known about feeding strategies of nocturnal benthivores, mainly because observers have difficulty watching these events under low illumination. Whereas visual, and in some cases tactile, cues appear to be most important for diurnal feeding on invertebrates, chemical cues likely play a significant role in night feeding.

## 2.4.2 Zooplanktivores

Two general mechanisms exist among fishes for capturing zooplankton: *filter-feeding*, where plankters are strained from the water by the gill raker apparatus as the fish swims forward with its mouth open; and *particulate feeding*, in which organisms are picked individually or a few at a time from the water by biting or sucking movements. The method used depends on the relative size of predator and prey, and on the relative abundance of different types of prey. For example, aquarium-held northern anchovies (*Engraulis mordax*) fed on brine shrimp (*Artemia salina*) nauplii about 0.6 mm long by filtering, and on *A. salina* adults (mean length 3.7 mm) by snapping at individuals (Leong and O'Connell, 1969). Film analysis showed that while filter-feeding the fish opened their mouths less often, kept them open for longer periods, and changed direction less often than while particulate feeding (Table 2.1). Increased turning rate was typical of the feeding "frenzy" that occurred as the anchovies broke up their schooling formation and turned rapidly and frequently while snapping at recently introduced *Artemia* adults. The "frenzy" did not occur when the same fish were given large quantities of nauplii. It appears to result from each fish visually fixating and then snapping at individual plankters rather than maintaining position in a polarized school as they generally do while filter-feeding. The feeding "frenzy" probably often occurs

**Table 2.1.** Distinguishing features of filter-feeding and particulate feeding in *Engraulis mordax*. (After Leong and O'Connell, 1969)

| Type of feeding | No. of fish observed | Mouth opening rate (no./5 s) | Mouth opening duration (s) | | | Turning rate (no./5 s) |
|---|---|---|---|---|---|---|
| | | | Mean | Max | Min | |
| Snapping at *Artemia* adults | 6 | 9.4 | 0.15 | 0.25 | 0.09 | 9.7 |
| Filtering *Artemia* nauplii | 6 | 2.9 | 1.72 | 4.45 | 0.63 | 1.3 |

among schooling planktivorous fish when they encounter dense but patchy accumulations of prey.

In another study, O'Connell (1972) found that the time spent by *E. mordax* in the two types of feeding depended on the relative abundance of *Artemia* adults and nauplii. When adults were about 2% of the available biomass, the fish fed equally by filtering and snapping. When adults were over 7% of the biomass, almost all feeding was by snapping. This variation probably increases feeding efficiency, since unless nauplii are extremely dense the fish could ingest more food in the same time by concentrating on adult *Artemia*.

The alewife (*Alosa pseudoharengus*) has at least four distinct feeding modes, that which is used depending on the species and density of available prey (Janssen, 1976, 1978). Relatively weak-swimming organisms, such as cladocerans, cyclopoid copepods and amphipods are captured individually by *particulate feeding*. The alewife orients towards a single prey, opens its mouth and sucks in the prey. Stronger swimmers, such as calanoid copepods and *Mysis* are caught by a modified form of particulate feeding called *darting*. The fish approaches a single planter, coasts with the body slightly bent in an S-shape, then darts suddenly with mouth open and sucks in the prey. This technique overcomes the vigorous negative rheotaxis of these organisms. When weak-swimming prey, such as *Daphnia*, are present at high densities alewives capture them by *filter-feeding*, in which the fish opens its mouth wide for 0.5 to 2 s while swimming forward vigorously with strong tail beats. Finally, *gulping* is shown by relatively large alewives when prey density is high. In this mode, the fish opens and closes its mouth 2 to 3 times per s, but does not change speed or orient towards individual prey. Usually several planters are sucked in with each mouth-opening. Gulping is thus intermediate between filtering and particulate feeding. It is similar to the "gape and suck" technique used by many fishes to capture several planters at once (Liem and Osse, 1975). A varied feeding repertoire, such as that shown by alewives, must be an efficient strategy for any species that forages in a rich plankton fauna, enabling the fish to adjust its capture technique to suit the most common prey available.

### 2.4.2.1 Prey Selection

The preceding descriptions raise the question of the ability of zooplanktivorous fishes actively to select certain prey items from among a range of available types. This has been examined by several workers.

Some selection is mechanical, as in filter-feeders; organisms smaller than the mean gap in the filtering apparatus tend to escape from the forager. On the other hand, mouth size may be inappropriate for some classes of prey. A study of yellow perch (*Perca flavescens*) in a Manitoba lake showed that perch fry below about 18 mm total length had a higher proportion of small *Daphnia pulicaria* in their stomachs than was found in the accompanying plankton fauna (Wong and Ward, 1972). Larger fry did not show this apparent selectivity. In aquaria, the smaller fry attempted to capture all *Daphnia* but released the larger ones. Their mouth gape was too small to admit the larger plankters. Growth in mouth size was faster than growth in body size, and thus the larger perch fry were soon able to utilize the entire *Daphnia* population.

Evidence that yellow perch and rainbow trout (*Salmo gairdneri*) actively select certain species and sizes of prey from a range of zooplankters was presented by Galbraith (1967). Only *Daphnia* spp. were consumed, and mostly those over 1.3 mm in size, despite the wide range in sizes of fish examined and species of plankton present. The mean gill raker gap for both fish species was such that more small organisms should have been present in their stomachs if prey were being selected only by the gill raker filter system.

A study in Lago Maggiore, Italy, showed that *Coregonus, Alosa*, and *Alburnus* were clearly selecting their prey from among the available zooplankton (Brooks, 1968). Generally the larger plankters were consumed first, with the fish turning to smaller types when the larger ones became scarce. Brooks argued that the large size and jerky, continual locomotion of the largest plankters made them most conspicuous to the visually hunting, particulate feeding fishes.

Finally, in an experimental study of prey selection by bluegill sunfish (*Lepomis macrochirus*), Werner and Hall (1974) found that at low abundance of prey (*Daphnia magna*) the fish ate those of different sizes, as they were encountered. With greater prey abundance the fish selected larger ones. In addition to demonstrating the capacity of a fish to select among food items of different sizes, this study provides quantitative support for the theory that animals can be expected to forage in such a way as to optimize their use of time and energy. More data from fishes are needed to broaden empirical support for the rapidly developing theory of optimal foraging strategies (Schoener, 1971; Pyke et al., 1977).

## 2.4.3 Aerial Feeders

Carnivorous fishes often feed on organisms floating at the surface in shallow water. Much less common is the ability to capture prey that is flying or resting on a surface above the water. Salmonids do this by jumping out of the water to capture low-flying insects (Kalleberg, 1958). Some other fishes rise to the water surface, take aim while their eyes are completely submerged, then spit water at resting prey, and quickly engulf the organism when it is knocked onto the water surface. Best-known for this ability are the archerfishes (Toxotidae) (Lüling, 1969; Vierke, 1969, 1973). In an elegant study Dill (1977) analyzed, with the aid of high-speed photography, the ability of *Toxotes chatareus* to compensate both for refraction of light at the water surface and for the effects of gravity on the airborne water droplets during spitting. Whereas *T. jaculatrix* (Lüling, 1963) and *Colisa* spp. (Vierke, 1973) minimize refraction problems by spitting from nearly directly below their intended prey, *T. chatareus* spits from a variety of positions. It is nevertheless capable of accurate compensation for refraction although, not surprisingly, spitting accuracy decreases with increasing distance between fish and prey. Dill's data also showed that *T. chatareus* compensates for gravity, and is thereby successful at hitting prey at various heights above the water, not by varying the spitting velocity, but by adjusting the angle of spitting. The adjustments required for accurate spitting are apparently made by the fish during a brief period of binocular fixation of prey just before it pitches up into a steeper body angle and spits (Dill, 1977).

The ability to capture aerial prey by spitting from the water surface would seem to be an effective foraging strategy directed at individual insects resting on vegetation close to the water. A single jet of water, shot with force and accuracy, is less conspicuous and therefore less likely to alarm the prey than is the fish itself jumping from the water. On the other hand, salmonids generally jump to catch flying insects when large numbers of recently emerged adults are flying near the water. These swarms of insects are not frightened away by jumping fish, and are therefore an abundant food source during the brief interval before they reproduce and die.

## 2.4.4 Piscivores

The detection and capture of other fishes as food can be described within the framework of four basic hunting strategies: ambushing, luring, stalking, and chasing. In addition, some fish are parasitic on others, gaining nutrition from their hosts but not killing them in the process.

### 2.4.4.1 Ambush Hunting

Piscivores that hunt by ambush are mostly diurnal predators, depending for success on surprising their prey. They accomplish this by remaining inconspicuous near the substrate, either by hiding, staying motionless, camouflage, or by combinations of these (Curio, 1976). When a potential prey fish moves within striking distance the predator turns slowly so as to orient towards it, then dashes forward, seizes the prey in its mouth and either returns to its former hiding place or consumes it on the spot. Ambush hunters seldom chase their prey if the first strike misses, although they may bite several times in quick succession at an elusive target.

Marine fishes specialized for ambush hunting include some groupers (Serranidae) that rush out from hiding places to overwhelm passing prey, and lizardfishes (Synodontidae) and flatfishes (Bothidae) that lie on or partly buried in sandy bottoms, where their cryptic colouring affords camouflage (Hiatt and Strasburg, 1960; Hobson, 1968a, 1974). There is little documentation of ambush feeding by freshwater piscivores, although species such as pike (*Esox lucius*) that typically stalk their prey (see Sect.2.4.4.3), may also use the ambush technique to capture prey that swim near their resting places in shallow, weedy waters.

30

**Fig. 2.4.** Various lures apparently used by piscivorous fishes to attract prey. **A** *Lophius piscatorius*. (After Norman and Greenwood, 1975), **B** the bathypelagic *Chaenophryne parviconus*. (After Hart, 1973), **C** *Phrynelox scaber*. (After Wickler, 1968), **D** *Iracundus signifer*. (After Shallenberger and Madden, 1973)

## 2.4.4.2 Luring

A number of marine fishes in the order Lophiiformes, often called "anglerfishes", are believed to attract potential prey by using a movable "lure". This consists of various modifications of the dorsal fin (Fig. 2.4). Usually the first dorsal spine is extended and has a fleshy or frilly flap of tissue, the esca, at its tip. The entire structure is called the illicium. It varies greatly in length, thickness and degree of branching. The esca is also variable in size and shape, and in some deepwater species is luminescent. The illicium has been compared to a fishing line with bait, because the anglers are said to remain motionless while moving the lure near their mouth. When small fish or invertebrates approach to examine the possible food item, the anglerfish suddenly opens its mouth and engulfs its prey (Wickler, 1968; Idyll, 1971; Norman and Greenwood, 1975).

The deep-sea anglerfishes (Ceratioidei) are bathypelagic and, although they have incredible diversity in development of the illicium and esca and of associated photophores (Bertelsen, 1951; Idyll, 1971), direct evidence of their feeding strategies is unavailable because of the inaccessibility of their natural habitat. Shallow-water "anglers" have been observed feeding, and there is clear evidence that the esca is effective as a fishing lure. Chadwick (1929) observed the goosefish *Lophius piscatorius* feeding in an aquarium. While lying still on the bottom with its large mouth closed, the illicum was moved so that the esca jerked to and fro just above the mouth. When a young coal-fish (*Gadus virens*) approached and examined the esca the angler suddenly opened its mouth and sucked in the small fish. *Pherynelox scaber* has a pinkish, wormlike esca that is jerked about, coiled and uncoiled, in front of its mouth. This "bait" attracts worm-eating fishes that may be captured by *P. scaber* if they approach closely (Wickler, 1968).

Fishes of some other groups are also said to lure potential prey. For example, the scorpaenid *Iracundus signifer* has a spiny dorsal fin that when erect, and seen from the side, closely resembles a small fish, complete with mouth, eye and dorsal fin (Fig. 2.4D). When hungry, *I. signifer* spreads its spiny dorsal and snaps it back and forth sideways. The fin colours deepen and the single black "eye spot" with its white border becomes conspicuous. Fish that approach this "lure" are attacked (Shallenberger and Madden, 1973).

### 2.4.4.3 Stalking

Some piscivorous fishes hunt by slow, stealthy approach to within striking distance of their prey, then suddenly lunging and attacking before the prey can escape. The distinction between this strategy and ambush hunting is that stalkers actively approach their prey. Successful stalking depends on slow, gradual approach by the predator that fails to elicit escape responses until it is too late.

A good example of this technique is provided by the northern pike (*Esox lucius*). The typical sequence of events is: fixating the prey with the eyes, turning and slowly moving towards the prey, leaping forward, snapping and finally swallowing it (Hoogland et al., 1957; Neill and Cullen, 1974). The fixating, turning, and approaching stages have been called "axial tracking" by Nursall (1973). Only the pectoral and dorsal fins appear to be used for the slow forward progression. Before striking, the pike pauses briefly and bends its tail to one side, so that tail and body together form a shallow S-bend. Then with a single powerful beat of the tail it leaps forward and grabs the prey in its mouth. The posteriorly pointing teeth prevent escape of the prey, which is swallowed directly and quickly if small, but is turned around and swallowed head-first if larger. The turning is done by a series of brief, vigorous head-shakes, presumably with the grip of the teeth loosening somewhat with each shake.

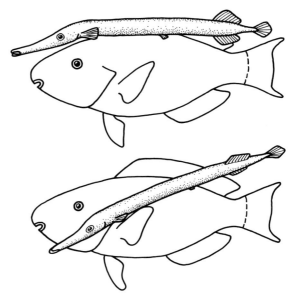

**Fig. 2.5.** Trumpetfish (*Aulostomus*) using a parrotfish (Scaridae) as cover while approaching prey. (After Eibl-Eibesfeldt, 1955)

Other species that regularly hunt by stalking much like the pike are the longnose gar *Lepisosteus osteus* (Foster, 1973), the mudminnow *Umbra limi* (J. Hawkins, pers. comm.) and the Lake Malawi cichlid *Haplochromis compressiceps*. The latter species approaches potential prey slowly from a hiding-place among weeds, pitched in a head-down orientation. It has an extremely laterally compressed body, and the narrow body outline, as seen from the front, apparently allows the slowly moving predator to approach within striking distance without alarming the prey (Fryer, 1959; Wickler, 1966a).

Some stalking predators have disruptive colouration in the form of a stripe running from the snout back along the mid-dorsal line to the dorsal fin. This stripe is coloured differently from the body on either side, thus producing the so-called "split-head" colour pattern that may well allow the hunter to approach its prey undetected (Barlow, 1967).

Even more interesting, from the ethological point of view, is the use of other fishes as cover by stalking predators. The long, slender-bodied trumpetfishes (Aulostomidae) and cornetfishes (Fistulariidae) sometimes swim close beside another fish of different body shape (such as a scarid, labrid or acanthurid), then suddenly dart out and attack nearby prey (Fig. 2.5). Apparently the close association of the two fish allows the predator to avoid detection before it attacks (Eibl-Eibesfeldt, 1955; Hobson, 1968a, 1974; Collette and Talbot, 1972). Single *Aulostomus* have also been seen swimming with a school of smaller heterospecifics, apparently using them as cover from which to dart at prey (Fig. 2.6). *A. maculatus* has several different colour phases of the head (Randall, 1968), and when one is associated with a school, its head-colour usually corresponds to that of the accompanying fish. For example, at the Grenadine Islands *A. maculatus* were seen

33

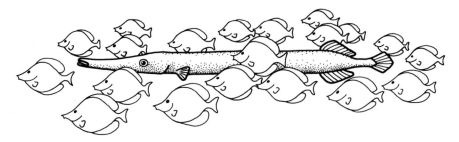

**Fig. 2.6.** Trumpetfish swimming in a school of surgeonfish (*Zebrasoma*). (After Eibl-Eibesfeldt, 1955)

in the purple colour phase (i.e., upper half of the head purple) while swimming in schools of the blue to purplish-coloured blue tang (*Acanthurus coeruleus*), and at Discovery Bay, Jamaica, *A. maculatus* with bright blue heads were seen swimming in schools of blue chromis (*Chromis cyanea*) (Kaufman, 1976). In the latter situation the trumpetfish occasionally struck at and captured one of the *C. cyanea* without eliciting the typical escape response of the school. Thus the predator's head colour, and its habit of associating closely with small schooling fishes, allow it to use the other fish as cover while approaching prey, or to prey on the schooling "host" species itself.

### 2.4.4.4 Chasing

Many piscivorous fish feed by chasing, overtaking, capturing, and ingesting individual fish. The prey is swallowed whole or in pieces, depending on its size relative to the predator. The surprise element is less important in this strategy than in ambushing, luring, or stalking. The chasing predator is successful because of faster, more efficient swimming and manoeuvering, or because the prey is trapped or becomes exhausted. Feeding by chase and capture is the commonest strategy for active, pelagic fishes that feed in open water, and is also used by benthic feeders that chase their prey from one hiding place to another.

Detection of and approach to prey by hunting piscivores are mediated by different sensory modalities, depending on the habitat and the cues produced by potential prey. Acoustic cues can be useful over long distances, in any direction. Chemical cues can also be effective at a distance, but in flowing water will only be detected downstream. Visual cues require appropriate illumination and proximity of predator to prey. The relative importance of the sensory modalities in eliciting feeding by piscivorous fish has been studied by several workers. Much effort has centred on sharks, because of the strong desire to learn how to reduce the incidence of shark attacks on humans (Gilbert, 1963).

The speed with which sharks often approach injured fish and a variety of underwater sounds, suggests they are highly sensitive to chemical and acoustic signals (Eibl-Eibesfeldt and Hass, 1959). Hobson (1963) compared the feeding responses of free-ranging sharks at Enewetak Atoll (Marshall Islands) to stimuli from dead, injured, or distressed prey. Grey sharks (*Carcharhinus menisorrah*) responded to uninjured, but tethered, struggling bait fish of several species by rapid and direct approach from downstream, where chemical and acoustic cues were probably the only ones detected. *C. menisorrah* responded similarly to colourless extracts from the flesh of groupers (Serranidae), approaching quickly from downstream when there was a noticeable tidal current flowing, but not at slack water. This supported Hobson's concept that sharks can follow an "olfactory corridor" consisting of a narrow trail of chemical cues carried downstream from a localized source. The role of vision in enabling a shark to distinguish edible from inedible prey was shown by presenting a piece of fish and a block of wood of the same size as the bait. When *C. menisorrah* approached, presumably in response to chemical cues from the fish, they struck at both baits indiscriminately. However, the wooden block was quickly rejected, while the fish bait was carried away. The baits were distinguished presumably on the basis of chemical and tactile stimulation of the mouth (Hobson, 1963).

Another demonstration of the key role of chemical cues in shark feeding was made by Tester (1963). Similar results were obtained with *C. menisorrah* and the blacktip shark (*C. melanopterus*), and three prey species (a grouper, *Epinephelus*; mullet, Mugilidae, and surgeon fish, Acanthuridae). When water from a small tank holding quiescent prey entered a large shark-holding tank the frequency of turns by sharks increased over control periods, but quickly waned. When prey fish were disturbed, by moving a stick about in their container, the sharks responded with greatly increased activity, including biting at the siphon leading from the prey tank. The specific source and identity of the stimulating material was not determined, but Tester (1963) suggested that under natural conditions sharks are habituated to chemical cues from small, potential prey species nearby as long as these fish are undisturbed. But if they are alarmed in any way, or if they rub against each other or a hard substrate, they may give off new chemical signals that elicit hunting responses in nearby sharks.

Sharks often gather so quickly in areas where explosions have occurred, that they must have been attracted by the sound vibrations (Eibl-Eibesfeldt and Hass, 1959; Hobson, 1963; Cousteau and Cousteau, 1970). However, the broadcasting of pure tones from 100 to 1000 Hz, and of mixed frequencies produced by wood striking against wood, rock against rock, and human shouts, elicited no observable responses from *C. menisorrah* or whitetip sharks (*Triaenodon obesus*) (Hobson, 1963). On the other hand, free-ranging silky sharks (*C. falciformis*) were attracted to sounds of from 25 to 1000 Hz, produced by a white noise generator (Myrberg, Ha, Walewski and Banbury, 1972). In an attempt to use more biologically meaningful sounds in a study of shark attraction, Banner (1972) recorded responses of young lemon sharks (*Negaprion brevirostris*) to broadcasts of sounds produced by several species of fish when swimming, jumping, feeding, and struggling, and to vocalizations from catfish (*Galeichthys*). Some of these sounds attracted sharks, others did not, and Banner (1972) concluded that lemon sharks use some sounds produced by potential prey fish to initiate searching behaviour and to locate prey.

The discovery that the shark *Scyliorhinus canicula* is highly sensitive to weak electric fields (Dijkgraaf and Kalmijn, 1963) opened up the possibility that sharks can use their electro-sensitivity to detect prey. In a series of experiments Kalmijn (1971) showed conclusively that *S. canicula* (as well as the ray *Raja clavata*) can detect a flatfish (*Pleuronectes platessa*) that is visually hidden on a sandy bottom by responding to the weak bioelectric field produced by the flatfish (Fig. 2.7). These

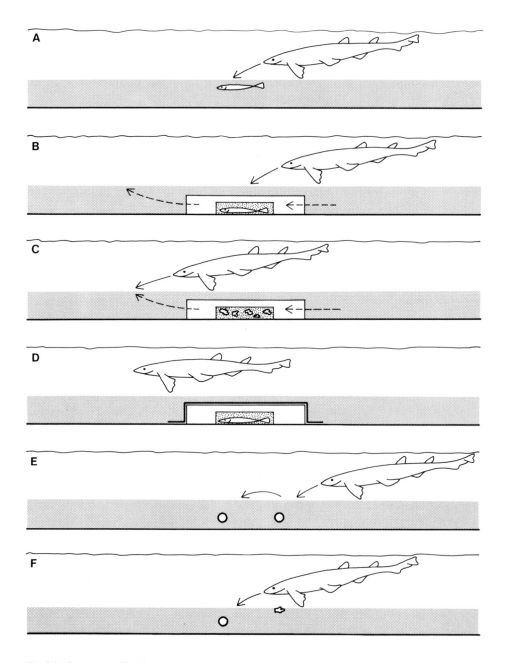

**Fig. 2.7.** Summary of feeding responses of *Scyliorhinus canicula* to various objects buried in sand. *Solid arrows*, responses of shark; *broken arrows*, flow of sea water through agar chamber. **A** plaice (*Pleuronectes*) under sand, **B** plaice in agar chamber, **C** pieces of whiting (*Gadus*) in chamber, **D** plaice in chamber covered with plastic film, **E** electrodes producing electric field, **F** piece of whiting and one electrode. (After Kalmijn, 1971)

studies established the existence of a functional electric sense, mediated by the electrosensitive ampullae of *Lorenzini*, in *S. canicula* and *R. clavata*. Presumably in nature this sensitivity enables the predators to detect prey that is hidden in the substrate, and not detectable through other senses. See Kalmijn (1974) for a review of the detection of electric fields by aquatic animals.

In general, these experimental studies of shark feeding have shown that search behaviour is most likely initiated by chemical and acoustic cues, that visual, chemical, and in some cases electrical cues are important in mediating approach and attack behaviour, and that chemical and probably tactile cues are used to distinguish edible from inedible material in the mouth.

Studies of feeding behaviour in piscivorous teleosts strongly suggest that vision plays the key role in detection, chase and capture. The most direct evidence is the observation that predatory fishes orient towards and visually fixate on moving prey, maintaining this orientation even while chasing prey that are fleeing along an erratic path (Deelder, 1951; Hoogland et al., 1957; Beukema, 1963; Olla et al., 1970). Also predators frequently attack only those individuals in a school that swim erratically or those that have become separated from the school (Hobson, 1968a). The tendency to attack strays and erratic swimmers is a common feature of predatory behaviour in general (Curio, 1976). It is presumably based on the increased conspicuousness of the unusual or separated fish, and possibly also on the decreased capacity of the erratic swimmer to avoid capture. Herting and Witt (1967) found that centrarchid fishes (*Lepomis macrochirus* and *Micropterus salmoides*) that were injured or stressed from handling, or emaciated from lack of food, were captured more readily by bowfin (*Amia calva*) than were controls. The stressed fish swam sluggishly and made slower and less frequent avoidance responses to attacking predators than did control fish. In another study pumpkinseed sunfish (*Lepomis gibbosus*) presented with four identical models of prey fish struck more often at the model moved along an erratic, wobbly course than at those moved along a straight line (Gandolfi et al., 1968).

A clear example of the vulnerability of disoriented fish in a school was provided by Hobson (1968a). Flatiron herring (*Harengula thrissina*) in the Gulf of California stay close to shore in large, quiescent schools by day. Shortly after sunset they move offshore to nocturnal feeding grounds. Hobson observed this offshore movement from a hill overlooking the beach. As the leading edge of the school passed over a sharp demarcation between rocky and sandy bottom these fish turned and swam back into the mass of the school. "The result was a localized breakdown in the structure of the school and a point of mass confusion. Large predators were accompanying the herring offshore, and...the surface of the water...became a froth of activity as the predators immediately converged on the disoriented segment of the school" (Hobson, 1968a, p. 55).

The increased vulnerability to predation of individual fish that become separated from a school, even if their swimming remains normal, has also been observed in nature. For example, rainbow trout (*Salmo gairdneri*) feed on schools of redside shiners (*Richardsonius balteatus*) by swimming close to the periphery of a tightly packed school, periodically charging and trying to catch individual shiners that become separated from the rest (Crossman, 1959). Marine piscivores often hover near schools of smaller fish, occasionally charging at the school and picking off individuals that become separated from the others in the confusion surrounding the attack (Eibl-Eibesfeldt, 1962).

These and other observations of piscivore attacks on those individuals within schools of prey fishes that are "confused", swim erratically, or stray from the main body of the school, emphasize the importance of visual cues in mediating successful predation. They also provide evidence on the effectiveness of schooling as an anti-predator strategy (see Chap.3).

### 2.4.4.5 Parasites

Few fishes are true parasites, in the sense of being structurally, physiologically and behaviourally dependent on a host organism for food and survival, but usually not killing the host. Of those that are, the ceratioid anglerfishes are perhaps the most spectacular (Bertelsen, 1951). There is extreme sexual dimorphism in many species, with the much smaller male becoming permanently attached to the female after metamorphosing from a free-living larval stage. The blood vessels of the two fish are apparently in contact, and the male thus obtains nourishment from the female (Norman and Greenwood, 1975). However, nothing is known of the behavioural interactions between the sexes before and during the attachment phase. A number of other fishes are specialized for feeding on parts of living fish (scales, fins, blood, other tissues). The best-known of these are the parasitic species of lampreys.

*(a) Lampreys (Petromyzontidae).* The parasitic lampreys feed on a variety of fish species, and some are known to attack whales (Pike, 1951). Detection and approach to prey are mediated by vision and a highly sensitive olfactory sense (Kleerekoper and Mogensen, 1963; Hardisty and Potter, 1971). The sea lamprey, *Petromyzon marinus*, has been studied in most detail (Lennon, 1954). Its circular oral disc is closed while the animal is free-swimming, but when attacking the head is raised, the oral disc opened and contact is made. A strong sucking action of the bucco-pharyngeal region, together with the many buccal teeth, maintain the grip. The tongue with its sharp anterior edge cuts through the host's skin, and blood and other tissues are sucked from the wound. Secretions from the large buccal glands have anticoagulant, haemolytic and cytolytic properties that modify the host tissues and make them easier to extract.

Lampreys do not stay attached for long to a dead or dying fish, and therefore the duration of attachment depends on the size and durability of the host as well as the feeding intensity and size of the lamprey. This may help to explain the tendency for *P. marinus* to attack the larger specimens of four different host species in aquarium tests (Farmer and Beamish, 1973). The larger the host fish the longer the lamprey can feed without killing it. Despite this possibility, the rapid decline in numbers of several fishes of the upper Great Lakes of North America after *P. marinus* had gained access by shipping canals (Smith, 1971) is evidence that parasitic lampreys can seriously damage and kill their hosts.

*(b) Catfishes (Trichomycteridae and Cetopsidae).* These Amazonian fishes, known locally as candirú, have an awesome reputation as parasites of other fish, and occasionally of man. Several species are known to enter the gill chambers of larger fish and feed on their blood. The evidence is from specimens taken from the gills of fish captured in the field and finding their stomachs gorged with blood (Roberts,

1972), and from observations of *Vandellia* sp. entering the gills of goldfish, feeding on their blood and then dropping out (Kelley and Atz, 1964). Presumably the greater the disparity in size between host and parasite the less likely is the former to suffer severely from such attacks.

Candirú attacks on humans can be far more devastating. The fish have the reputation of being strongly attracted to human urine, and very small individuals are said to have entered the urethra of both men and women, from which they were removed, if at all, with great difficulty and pain to the "host". These reports are detailed enough that they likely have a factual basis, even if the incidence of attacks on humans is low (Gudger, 1930a,b).

*(c) Scale-Eaters.*Feeding on scales taken from other fish is a highly specialized trophic adaptation found in some African cichlids, South American characoids, and a few other fishes (Major, 1973; Mok, 1978). The evidence for the habit consists of finding large numbers of fish scales and little else in the predator's stomachs and, in a few cases, of observing scale-eating by aquarium-held fish. For example, when a single *Leporinus* sp. was placed in an aquarium with several smaller *Exodon paradoxus* (both species are Amazonian characoids), the latter took "...turns making extremely rapid circular stabbing motions against its side, always striking towards the free margin of the scales, and removing a single scale at about every other strike. The scales were swallowed directly" (Roberts, 1970, p. 389). Several cichlids of Lakes Tanganyika, Victoria and Malawi, consume scales regularly. Two scale-removal techniques are used. One is to bite or scrape at one or a few scales and thus pull them off the attacked fish. The other is to grasp the tail of another fish in the mouth and rasp away large numbers of the very small scales that occur on both sides of the caudal fin (Fryer and Iles, 1972).

All scale-eaters have highly specialized dentition that facilitates scale removal (Liem and Stewart, 1976). It appears that the attacked fish are not seriously injured by these encounters, although repeated loss of scales, especially from the caudal fin, must impair their capacity to swim.

*(d) Fin-Biters.*Some fishes feed regularly on pieces they bite from the fins of other fish. Several African characoids in the family Ichthyboridae belong to this group; the stomachs of adults usually contain nothing but pieces of fish fins. The technique used by one such species, *Belonophago hutsebauti*, was observed in an aquarium. "The fish 'hovered' in horizontal position just below some floating grass...As soon as some other fish swam within reach, it darted out, grabbed hold of a fin between its long jaws and snipped off part with a quick twist..." (Matthes, 1961, p. 79).

Fin-eating also occurs among marine fishes. The two blenniids *Runula albolinea* and *Aspidontus taeniatus* have been observed by divers biting pieces from the fins and also the skin of larger fishes (Eibl-Eibesfeldt, 1955, 1959). *A. taeniatus* appears to be an effective mimic, in size, colour, and behaviour of the "cleaner-fish" *Labroides dimidiatus*. It approaches a much larger fish in the same manner as *L. dimidiatus*, then suddenly darts forward, bites a piece from the fin or skin of the "host" and retreats (Eibl-Eibesfeldt, 1959; Randall and Randall, 1960).

In all of these cases, the attacked fish do not seem to suffer unduly. However, repeated damage to fins by many such attacks could gradually weaken a fish, and eventually reduce its ability to escape predation.

## 2.5 Group Feeding

Many fishes that are loosely aggregated or even scattered when not feeding form tighter aggregations or schools as they begin to feed (Hobson, 1968a). Foraging threespine sticklebacks (*Gasterosteus aculeatus*) quickly converge and search for food around one individual that pitches into a head-down position and feeds on newly discovered prey (Keenleyside, 1955). On coral reefs, where species diversity is high, fish that begin feeding are often joined by others, of the same and of different species, in group exploitation of a localized food resource (Fishelson, 1977). These observations strongly suggest there are advantages to group, as opposed to solitary foraging. However, an explanation of the advantage might well depend on the type of prey organism being utilized. The relation between foraging success and social behaviour of the foragers would not be expected to be the same for prey as variable as plankton, fish, and benthos. Yet social feeding on these three categories of food occurs.

### 2.5.1 Planktivores

Planktivorous fishes often feed in tight schools, even though one would think that more widely spaced foragers would have greater success because of reduced competition. In fact, Eggers (1976) showed on theoretical grounds that planktivorous fishes whose prey is randomly distributed will decrease their prey encounter rate as the mean inter-fish distance decreases. This is caused by increasing overlap of visual fields as the fish school more closely, and also by removal of plankton by the preceding members of a school, leading to reduction in numbers of prey from front to end of the school. In view of these relationships, the observed schooling of many planktivores may have the following explanations. First, a random distribution of plankters is not likely realistic in most cases. The effect of schooling behaviour on feeding success when plankton is patchily distributed will then depend on the nature of that patchiness (Eggers, 1976). Second, foraging planktivores probably school, at least in part, as a response to potential or real attacks on them by their predators, especially when their feeding activities take them into open water, away from sheltered areas where they stay while not feeding. Schooling is characteristic of many shallow water species while feeding in the water column above the substrate (Hobson, 1968a).

### 2.5.2 Piscivores

Piscivorous fishes also often feed in groups, and the possibility that joint action is a more effective hunting strategy than individual foraging has been considered by a number of authors. A simple form of group hunting is sometimes used by perch (*Perca flavescens*). They form a loosely-integrated, non-polarized group within which each fish searches independently for food. If one strikes at a prey fish others quickly approach and join in the attack. A single prey may be struck at repeatedly by different perch and finally eaten by one (Nursall, 1973). After observing *P. fluviatilis* foraging in a similar manner on roach (*Rutilus rutilus*), Deelder (1951) concluded that a single roach may be able to avoid one perch, but when several perch attack, the effort to escape from one quickly puts the roach into the attack zone of another. This type of joint attack on prey is probably common among piscivores (Hobson, 1968a, 1974).

Many fast-swimming, pelagic, marine piscivores forage in groups or schools, and the suggestion has been made that they cooperate in hunting. However, most of these species feed on much smaller, densely schooling fishes, with each predator apparently acting independently. Direct evidence of cooperative, joint hunting is rare. Eibl-Eibesfeldt (1962) saw a school of horse mackerel (*Caranx adscensionis*) continually circling around a school of snappers (*Caesio* sp.), moving them closer together, toward the water surface and away from the reef. Occasionally a single *Caesio* darted away from the school and was snapped up immediately by one of the circling predators. The same author observed a group of sharks (*Carcharhinus*) drive a school of small fish close to the shore of an enclosed bay, and then feed on individuals in the tightly packed group. Coles described a school of 100 or more sand sharks (Odontaspididae) driving a school of bluefish (*Pomatomus salatrix*) "...into a solid mass in shallow water, and then at the same instant the entire school of sharks dashed in on the blue-fish" (Coles, 1915, p. 91). Hiatt and Brock (1948) saw three black skipjack (*Euthynnus yaito*) actively herding a school of scad (*Decapterus*) in shallow water, preventing them from moving to deeper water. One laggard was picked off and eaten by a skipjack, but no other feeding was seen. However, Hobson (1968a) suggested that this may not have been a case of active herding by the predators, but rather that the scad were avoiding capture by staying in a shallow zone where they were relatively secure. The fact that only a single scad was captured during a three-hour observation period supports Hobson's interpretation.

It is clear that unequivocal evidence of collaborative, joint hunting by fishes in the manner of some carnivorous mammals (Curio, 1976) is rare. The undoubted group feeding of many piscivores seems rather to be a matter of fish individually attacking schools of prey at points of confusion or disruption of smooth swimming (Hobson, 1968a; Radakov, 1973). Each such attack, in itself, may cause disruption of the school, leading to more cases of disorientation, and so on. Perhaps the simplest explanation of the effectiveness of group hunting is that predators acting together can more easily restrict the movements of a school of prey fish than when they are alone. This allows the predators to maintain contact over long periods with their prey and feed to satiation without extensive preliminary hunting.


## 2.5.3 Benthivores

Some fishes that move about close to the substrate foraging on benthos often travel in schools. Well-known examples from coral reefs are parrotfishes (Scaridae), surgeonfishes (Acanthuridae), and goatfishes (Mullidae) (Hiatt and Strasburg, 1960; Hobson, 1974), and from freshwaters are cyprinids and catostomids (Scott and Crossman, 1973). Recently the suggestion has been made that foraging in groups is a strategy that gives these fish access to feeding areas from which single fish are excluded by the aggressive behaviour of resident, territory-holding fish.

**Table 2.2.** Species that feed in other fish's territories while travelling in schools. Species in brackets are associates that occasionally join intruding schools

| Intruder species | Resident species | Common diet | Locality observed | Authority |
|---|---|---|---|---|
| *Acanthurus triostegus* | *A. leucosternon* and *A. lineatus* | Algae | Addu Atoll, Indian Oc. | Eibl-Eibesfeldt, 1962 |
| *A. triostegus* | *A. achilles* and *A. nigrofuscus* | Algae | Johnston and Hawaiian Islands | Jones, 1968 |
| *A. triostegus* | *A. glaucopareius* | Algae | Tuamotu Archipelago, Pacific Oc. | Randall, 1961a |
| *A. triostegus* | *A. nigrofuscus* | Algae | Hawaii | Barlow, 1974b |
| (*A. nigroris*) | *A. nigrofuscus* | Algae | Hawaii | Barlow, 1974b |
| *A. sohal* | *Pomacentrus lividus* | Algae | Red Sea | Vine, 1974 |
| *A. coeruleus* | *Eupomacentrus* spp. | Algae | Caribbean Sea | Alevizon, 1976 |
| (*Scarus coelestinus*) | *Eupomacentrus* spp. | Algae | Caribbean Sea | Alevizon, 1976 |
| *Scarus croicensis* | *E. planifrons* and *S. croicensis* | Algae | San Blas Is., Panama | Robertson et al., 1976 |
| (*A. coeruleus, Chaetodon capistratus, C. ocellatus, C. striatus*) | *E. planifrons* and *S. croicensis* | Benthic invertebrates | San Blas Is., Panama | Robertson et al., 1976 |

The phenomenon has been observed primarily in coral reef ecosystems, where areas rich in living coral and associated organisms provide food for a wide variety of fishes. Much of this substrate may be occupied by territorial benthivores, especially pomacentrids. The residents are usually able to keep one or a few fish away, but if the territory is invaded by a school of actively feeding benthivores, the resident is swamped, and the intruders are able to feed, at least temporarily. Examples of species that have been observed to enter other fish's territories and feed when travelling in schools, but not as individuals, are listed in Table 2.2. The study of Robertson et al. (1976) provided quantitative data showing that *Scarus croicensis*, which forages either alone or in schools, benefits from schooling. The feeding rate was higher and the number of attacks per intruder by territory residents was lower when *S. croicensis* foraged in schools than when alone.

Another feature of group invasion of territories is that the intruding schools sometimes include a few members of other species (Table 2.2). The associates apparently benefit in the same way as the predominant members of the school. In the study of Robertson et al. (1976) at least four other species joined the intruding parrotfish schools. One of these (*Acanthurus coeruleus*) has much the same algal diet as *S. croicensis*, but the three chaetondontid associate species are benthic carnivores.

## 2.6 General Comments on Feeding Behaviour

The wide range of feeding strategies among fishes outlined in this chapter reflects the incredible diversity of aquatic organisms that are potentially available as fish food. Some of this organic material, for example benthic algae and higher plants, requires no special capture techniques; to utilize it as food fishes need only the appropriate dentition and digestive apparatus. Sessile invertebrates, such as sponges, coelenterates and molluscs have a range of defense mechanisms (including spines, nematocysts and shells) that require specialized structures and techniques on the part of fishes that feed on them. Motile organisms, both invertebrates and fish, are captured by fishes using methods ranging from straightforward chase and capture, to highly specialized strategies involving both behavioural and structural adaptations to overcome their prey's defense mechanisms.

All of these feeding methods have evolved in natural environments where certain recurring events influence the activity of fishes and of their animal prey. The most obvious of these, for all but deep-water species, is the 24-h cycle in illumination. Studies by divers have yielded important information on the interrelation between feeding behaviour and light. The work of Hobson at several Pacific Ocean sites is an excellent example (Hobson, 1968a, 1972, 1974).

Fishes of Hawaiian coral reefs and of coastal Gulf of California can be classed as nocturnal, diurnal, or crepuscular feeders. Many nocturnal feeders are planktivores, taking the larger types of crustaceans which are themselves most active in the water column at night. Conspicuous examples of such fishes are the holocentrids (squirrelfishes) and apogonids (cardinalfishes). Crepuscular feeders include large, active piscivores, such as some carangids (jacks and roosterfish) that capture their prey by charging and chasing. They capitalize on the increased vulnerability of many smaller schooling fishes that are in transition between feeding and resting areas during the short twilight period.

The above fishes are considered to be relatively generalized in their feeding behaviour. In contrast, the diurnally active feeders include many species that are specialized in structure, behaviour, or both, for foraging on certain kinds of prey. For example, diurnal piscivores include species that use ambush or stalking techniques to surprise their prey. Presumably these are more effective than straightforward charging attacks when illumination levels are high and prey could use evasive tactics against clearly visible predators. Many other fishes use specialized hunting methods and mouth parts to discover and capture small benthic prey, primarily invertebrates, that hide on the substrate. Diurnal feeders also include fish that specialize on sessile invertebrates (see above) and on vegetation. Well-known coral reef families, such as the Chaetodontidae and Pomacentridae, include some species that prey on benthic invertebrates or small types of diurnally active plankton, others that are strictly herbivores, and still others that eat both invertebrates and vegetation. These illustrate the adaptive radiation in feeding behaviour that has occurred among related fishes in the rich, complex coral reef ecosystem (Hobson, 1974).

Finally, it should be kept in mind that feeding strategies have evolved not only in response to variety in form of potential prey, but also in relation to predator-avoidance behaviour. All but the largest, fastest-swimming fishes are themselves subject to predation, and to be successful a species must feed in such a way and at such a time that its own vulnerability is minimized. The next chapter considers anti-predator devices among fishes.

Chapter 3

# Anti-Predator Behaviour

If fish are to survive and reproduce they must not only acquire adequate food, but must avoid becoming the food of other animals. Many different mechanisms exist among fishes for countering the feeding strategies of their predators. These can be grouped into two main categories, labelled primary and secondary defence mechanisms by Edmunds (1974). Primary mechanisms are those that reduce the probability of a fish being detected, or if detected, of being treated as potential food. They operate whether or not a predator is in the vicinity. Secondary mechanisms are those that reduce the probability of a fish being caught, once an encounter with a predator has begun.

## 3.1 Primary Anti-Predator Mechanisms

The most widespread strategies used to avoid detection by a predator are hiding and the use of camouflage. Clearly these overlap; an individual that is well camouflaged because it blends into its surroundings and is therefore inconspicuous, is also in a sense hidden; but many fish avoid detection without the use of camouflage. They hide by remaining in locations where they do not emit visual, acoustic, or other cues that could be detected by hunting predators.

### 3.1.1 Hiding

Fish will use virtually any space or enclosure large enough to accommodate them as a hiding place. Thus they hide inside burrows or tubes excavated in the substrate, under rocks, leaves, or other litter (including man-made debris), inside empty shells, in spaces within branching corals, in dense vegetation, and even among the tentacles or spines of other animals. The most common hiding species are small, benthic forms such as many of the gobies (Gobiidae), gunnels (Pholididae) and pricklebacks (Stichaeidae) in the sea, and some darters (Percidae) and sculpins (Cottidae) in freshwater. Some species remain in hiding most of the time; others use their shelter only during the inactive part of their 24-h cycle.

Some well-hidden fishes have been of special interest to behaviourists because they live in close association, often symbiotically, with one or more other species. For instance, some gobies occupy burrows excavated by shrimps (Magnus, 1967a; Karplus et al., 1972). The arrow goby (*Clevelandia ios*) lives in a large burrow shared by an echiuroid worm (*Urechis*), a crab (*Scleroplax*), a clam (*Cryptomya*)

44

and a scale-worm (*Hesperonoe*) (Gotto, 1969). The so-called "anemone-fishes" (*Amphiprion* spp.) are pomacentrids that live in permanent, close association with large sea anemones (Mariscal, 1972). Some apogonid cardinalfishes (*Siphamia*) live among the spines of large sea urchins (*Diadema*) (Eibl-Eibesfeldt, 1961; Fricke, 1970, 1973b), and the juveniles of several pelagic fishes swim among the tentacles of jellyfishes (Walford, 1958; Mansueti, 1963). One nomeid species (*Nomeus gronovii*) is called the man-of-war fish because it apparently spends its entire life swimming under the umbrella of the Portuguese man-of-war *Physalia pelagica* (Marshall, 1965). Perhaps the ultimate hiding place is used by the pearlfish *Carapus acus*. This long slender fish spends most of its time within the coelomic cavity of holothurian sea-cucumbers, occasionally entering and leaving via the anus and apparently feeding on the host's internal organs (Arnold, 1953; Trott, 1970).

In some of the above cases the sharp spines or nematocysts of the "host" may deter attacks on the associated fish, in which case the association should be considered a secondary defence mechanism. However, it often appears that the small fish remain well hidden, and thus avoid discovery by predators. Evidence that hiding is an effective primary defence mechanism is largely circumstantial. If a human observer has difficulty detecting a hidden fish, it is often assumed that natural predators will also have difficulty. An excellent reminder that this may not be so is the ability of some hunting sharks to discover flatfish that are completely hidden from view under sand, by detecting the extremely slight electrical output of the flatfish (see Sect.2.4.4.4).

## 3.1.2 Camouflage

Many fishes avoid detection by crypsis, that is, they are camouflaged to resemble their surroundings, and are thus inconspicuous (Edmunds, 1974). Among actively swimming pelagic fishes camouflage usually takes the form of countershading. More sedentary species have a wide range of structural and colour adaptations that make them cryptic. In both cases, locomotor behaviour and orientation are important parts of the camouflage.

### 3.1.2.1 Countershading

This is a principle of colouration in which the animal's dorsal surface is darkly pigmented, the sides become lighter ventrally and the ventral surface is lightest of all (Cott, 1940). This arrangement counters the pale dorsal and dark ventral surfaces that are created by light from above (as in upper layers of the water by daylight) on a uniformly coloured body (Fig. 3.1). Countershading makes pelagic aquatic animals appear inconspicuous from any direction, because whatever angle the body is viewed from, it appears similar to the background. Some pelagic fish, such as clupeids, are not only countershaded, but also have silvery reflecting material on the ventro-lateral surfaces. This increases the reflection of available light from the lower sides of the fish, making them even more inconspicuous when seen from a wide range of angles below (Edmunds, 1974). The presence of silvery, reflecting layers, however, makes the fish vulnerable to predation if they roll to one side, because the flashing reflections from the lower sides are then highly visible.

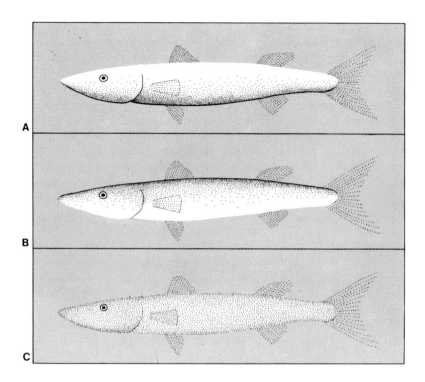

**Fig. 3.1.** The effects of countershading on visibility. **A** a uniformly coloured fish illuminated from above, **B** a countershaded fish uniformly illuminated from all sides, **C** a countershaded fish illuminated only from above. (After Cott, 1940)

Clupeids that swim erratically are often selectively attacked by predators, probably because of the brief loss of countershading (Hobson, 1968a).

Despite being countershaded, a pelagic fish seen directly from below, will appear as a silhouette against a lighter background, at least at depths to which light from the sky penetrates (Hobson, 1966). This has led to the suggestion that the function of ventrally placed photophores (common to many pelagic marine fish) may be to obliterate the fish's silhouette by approximately matching the background illumination, as seen from below (Clarke, 1963). Indirect support for this hypothesis comes from the fact that in most fishes with well-developed photophores, these are on the ventral part of the body, and the light they produce is directed ventrally (McAllister, 1967). Also, most of these fish occur at depths to which some light penetrates from the surface, thus creating the silhouette problem (Clarke, 1963).

One bioluminescent species, *Leiognathus equulus*, produces light as a diffuse glow over much of its ventral surface, and controls the amount of light emitted by a type of shutter around the single large light organ that contains luminous bacteria (Hastings, 1971). Specimens in captivity did not emit light in complete darkness, but did so briefly in response to a light stimulus. Such a system would be useful as an anti-predator mechanism if *L. equulus* were to emit light and thus disrupt its

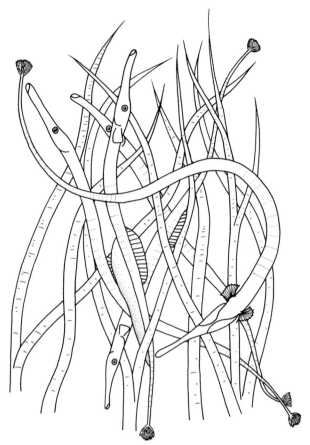

**Fig. 3.2.** A group of pipefishes (*Siphostoma*) holding position among eelgrass (*Zostera*). (After Marshall, 1965)

silhouette during daylight, but remain completely dark by night (Hastings, 1971). No direct evidence is available to support this anti-predator strategy, but the suggestion would be well worth exploring in more detail.

### 3.1.2.2 Resemblance to Surroundings

Many fishes that live close to the substrate, or among floating or rooted plants, are to some degree camouflaged by the similarity in colour between them and their background. However, some species are highly specialized in this regard, and through a combination of colour, morphology and behaviour are virtually indistinguishable from their immediate surroundings. The best-known examples are from the sea. The pipefishes (Syngnathidae) live among eelgrass (*Zostera*) and sea weeds in shallow, coastal waters. In *Zostera* beds they remain in a vertical position, their long slender bodies swaying back and forth with the leaves, and their colour closely matching that of the plants (Fig. 3.2). Associated with floating masses of Sargassum weed are a number of frogfishes (Antennariidae) whose

47

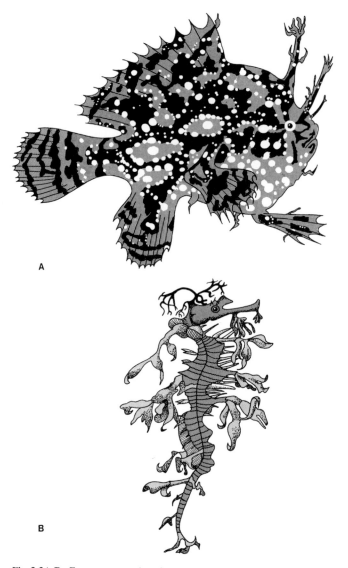

**Fig. 3.3A,B.** Extreme examples of camouflage. **A** the frogfish *Antennarius marmoratus;* **B** the seadragon *Phyllopteryx eques.* (After Cott, 1940)

blotchy, speckled, yellow, brown and white colour patterns and numerous fleshy protuberances give them a bizarre appearance (Fig. 3.3A) and make them virtually invisible among the weed beds. An even more extreme case is the syngnathid seadragon *Phyllopteryx eques.* Attached to its unusually shaped body is an incredible array of filaments, fronds, and spines, of different size, shape, and complexity (Fig. 3.3B). Authors are often reduced to describing this bizarre little animal as more like a complex mass of seaweed than a fish (Cott, 1940; Breder, 1946; Marshall, 1965).

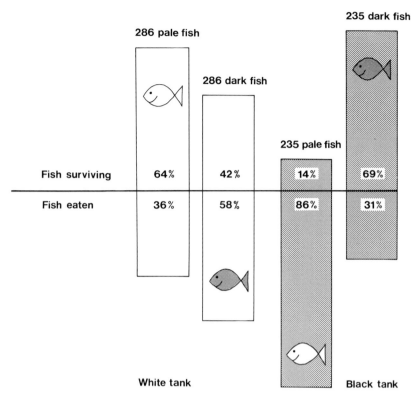

**Fig. 3.4.** Effects of dark and pale colouration in *Gambusia* on predation by penguins in black and white tanks. (After Edmunds, 1974, from Sumner, 1934)

A number of fishes, especially while juvenile, are said to resemble leaves, seed pods, twigs, or clumps of algae (Breder, 1946; Randall and Randall, 1960). To be concealed effectively from predators, as they often are from humans, the fish must resemble the plant material in shape, colour, and especially movement. Young *Lobotes surinamensis* provide a good example of this multiple similarity. In the shallow coastal waters of the Caribbean Sea juvenile *L. surinamensis* may be found floating near the surface close to dead and decaying leaves of the red mangrove (*Rhizophora mangle*) and the black mangrove (*Avicennia nitida*). These highly compressed little fish rest on one side, with head tilted slightly down, moving imperceptibly by slight movements of the transparent pectoral fins. Their yellowish-tan colour with many black spots is similar to that of decaying mangrove leaves. When leaves were placed in a tank with a single fish it moved toward them and stayed close by, as though "schooling" with the leaves (Breder, 1949). Similar camouflage through a close resemblance to fallen leaves has been reported for young *Lobotes* in forest pools of Ghana, where the fish are said to take on any of three different colour patterns found among the submerged leaves (Edmunds, 1974).

The ability to change the colour pattern of the dorsal surface to blend with the background is widespread among demersal fishes. Striking demonstrations of this

49

ability are found in the flatfishes (Pleuronectiformes). The pioneering investigations of Sumner (1911) illustrated the remarkable ability of these fish to camouflage themselves by matching their colour patterns to a wide range of background patterns. The photographs accompanying Sumner's paper are convincing evidence of this ability.

However, despite the wealth of observations on fish camouflage, there is little experimental evidence that fish are actually protected against predation by their camouflage. Sumner has provided some such evidence. In one experiment Galapagos penguins (*Spheniscus mendiculus*) were released into two large tanks, one painted white, the other black. Also released into each tank were equal numbers of dark and pale mosquitofish (*Gambusia patruelis*). The prey fish colours had been induced by holding them for 7 to 8 weeks in tanks painted white or black. The results were reasonably clear-cut (Fig. 3.4). In the white tank the darker fish were at somewhat of a disadvantage; in the black tank the paler fish were clearly more vulnerable (Sumner, 1934). In another experiment green sunfish (*Lepomis cyanellus*) were used as predators and *Gambusia* again as prey. The results were similar. Dark fish were more vulnerable in a pale grey tank, light fish in a black tank (Sumner, 1935). This work strongly suggests that predators will capture first those fishes that are relatively conspicuous in a uniform environment. Camouflage was an effective anti-predator mechanism in the experimental tanks, and presumably it can also be under natural conditions.

## 3.2 Secondary Anti-Predator Mechanisms

When an encounter between predator and prey begins, the prey animal takes action to avoid capture. The various forms of this action can be grouped into three general categories. A fish can: flee into a shelter; perform evasive action, independently or as a member of a group; or show aggressive defence to induce the predator to stop its attack. These will be discussed separately.

### 3.2.1 Flight into Shelter

Fish that stay near familiar burrows, caves, or crevices can avoid predators by rapid flight into shelter. Provided the predator cannot follow them, this should end the encounter. Many small coral reef fish provide typical examples. Individual pomacentrids that occupy territories or restricted home ranges on a reef dive for cover into the reef interstices when disturbed. Long-term residence at a location undoubtedly means that the fish are familiar with escape routes and shelters. Some schooling pomacentrids show the same flight response, but in a synchronized manner. For example, *Chromis cyanea* and *C. caerulea* are small, brightly coloured fishes that hover during the day in dense schools above coral heads, feeding on plankton. At night they settle into crevices among the coral. When alarmed by roving, diurnal predators, such as carangids and scombrids, they dive together into the coral and hide (Hartline et al., 1972; Fishelson et al., 1974; Hobson and Chess, 1978). In many cases the feeding schools contain several species; when alarmed they all dive together into shelter holes (Fishelson et al., 1974).

Some predator features provide a stronger stimulus for synchronized escape than others. In a study using ingeniously manipulated models in the natural habitat of *C. cyanea* at the U.S. Virgin Islands, Hurley and Hartline (1974) found that darker models were more effective than lighter ones, and larger models more effective than smaller. They also observed that fast-swimming natural predators consistently released the sudden, synchronous flight response, whereas the fish responded to slower-moving predators with slower and more independent movement to shelter. Thus, speed, size, and contrast with background were important features of predators that released synchronized flight of *C. cyanea*. These characteristics are probably associated with relatively dangerous predators that require a sudden and fast response if they are to be avoided.

Some fishes that live near the bottom do not have specific, familiar escape sites, but simply dive for cover under any available substrate materials. For example, mud minnows (*Umbra limi*) live in weedy areas, where the soft bottom is often covered with leaf litter and other debris (Scott and Crossman, 1973). When disturbed they dive into the substrate and remain motionless for long periods (Hawkins, 1974).

Other fish will quickly bury themselves in loose substrate by vigorous lateral body and tail movements when disturbed. This has been described for the killifish *Fundulus kansae* (Minckley and Klaassen, 1969) and *F. diaphanus* (Colgan, 1974). Small cichlids (*Aequidens* spp.) may bury themselves completely in fine gravel when being chased in an aquarium with a dip-net. The flatfish *Solea vulgaris* can cover itself quickly with sand by rocking and settling movements called "digging-in" by Kruuk (1963). The capacity for rapid self-burial, followed by keeping still for long periods, must be an effective escape mechanism, especially against visually hunting predators.


## 3.2.2 Individual Evasive Action

Fleeing, by rapid movement away from an approaching predator is probably the most common escape tactic used by animals (Edmunds, 1974). A single fish, in the absence of shelters, will often try to escape by swimming along an erratic, zig-zag path (Fig. 3.5). This has been called protean behaviour, i.e., "...that behaviour which is sufficiently unsystematic to prevent a reactor predicting in detail the positions or actions of the actor" (Humphries and Driver, 1970, p. 286). Unpredictable escape swimming is effective if it gives the prey animal time to find a safe shelter, or if the predator gives up and switches to another target. On the other hand, the onset of fatigue in the prey, or the presence of several predators, any of which can block escape routes and take up the chase, set limits to the effectiveness of protean escape reactions by a solitary fish.

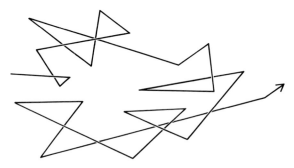

**Fig. 3.5.** The path followed by a stickleback (*Gasterosteus*) while fleeing from a fish-eating duck. (After Humphries and Driver, 1970)

### 3.2.3 Group Action

Many group-living fishes react to an approaching predator not by diving into shelter, but by clumping more closely together and swimming away in a polarized school. Threespine sticklebacks (*Gasterosteus aculeatus*) clearly show these changes in polarization and direction of movement when disturbed (Fig. 3.6). Further action depends on the predator's behaviour. If it moves slowly through the school without attacking, the prey fish may simply withdraw on all sides, forming an open space, or vacuole (Eibl-Eibesfeldt, 1962) around the predator as they stream around it, reforming into a single school after it has passed (Fig. 3.7). If the predator attacks, the prey often explode outward in all directions as each individual tries to avoid capture, then quickly draw together again. This "flash expansion" (Nursall, 1973) may be followed by a short period of agitation, during which the fish move about in a disorganized mass (Potts, 1970), but usually the polarized school quickly reforms. These changes in density, polarization and direction of movement of fish schools under attack have been documented for many species in nature by underwater observers (e.g., Eibl-Eibesfeldt, 1962; Hobson, 1968a; Potts, 1970).

The mechanisms by which schooling behaviour contributes to escape from predatory attacks have been the subject of much speculation, with few supporting data. Some frequent suggestions will be discussed briefly.

#### 3.2.3.1 Reduced Probability of Detection by Predators

Brock and Riffenburgh (1960) showed, on theoretical grounds, that fish are more likely to be detected by solitary predators if they are scattered than if the same number of prey are clustered in schools. This is because the frequency of predator–prey encounters is an inverse function of the number of schooled prey. However, this theoretical advantage of schooling may be offset by the tendency for some predators to remain close to large schools of prey, feeding occasionally, but not losing contact with the prey (Hobson, 1968a). In addition, many predatory fish hunt in schools. Therefore, realistic predictions about encounter frequencies must take into account the social organization of the prey species and its most common predators.

52

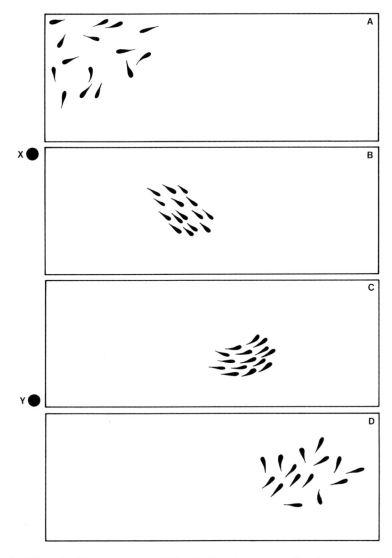

**Fig. 3.6.** A school of sticklebacks (*Gasterosteus*) in relation to disturbance. **A** undisturbed school, **B** observer arrives at *X*, **C** observer moves to *Y*, **D** observer leaves. (After Keenleyside, 1955)

## 3.2.3.2 Increased Probability of Detecting Predators

The suggestion has often been made that as the number of animals, and hence the number of peripheral sense organs, in a group increases, the speed of detection of approaching predators should increase (Bowen, 1931; Eibl-Eibesfeldt, 1962; Radakov, 1973). Therefore fishes should be able to take evasive action more quickly when in large than in small schools. This seems self-evident until one considers the case of very large schools. For instance, will 5,000,000 herring perceive an approaching predator faster than 500,000 or 50,000? The proportion of

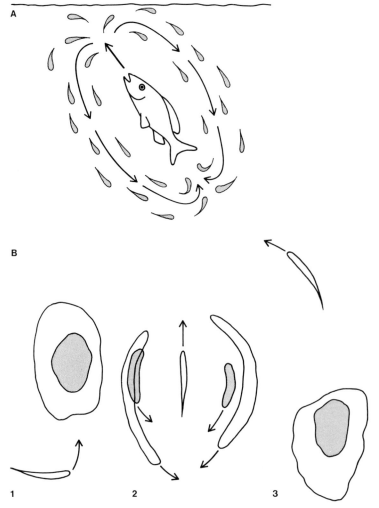

**Fig. 3.7.** Reactions of schooling prey fish to an approaching predator. **A** school of *Lutjanus monostigma* avoiding a predator (*L. bohar*). (After Potts, 1970). **B** schools of spottail shiners (*Notropis hudsonius*) and perch (*Perca flavescens*) (*stippled* and *clear areas* respectively) reacting to the passage of a pike (*Esox lucius*). (After Nursall, 1973)

individuals on the periphery of a school, where visual detection of predators first occurs, will decline as the total school size increases, in accordance with the decrease in the surface area/volume ratio of a sphere as the sphere increases in size. The protective value of massed receptor organs may be more evident in small to moderate-sized schools. Perhaps as long as all or most individuals in a group have equal probability of detecting an approaching predator, then larger groups will be warned more quickly. Once the school becomes so large that many individuals are in its interior, screened off from predator cues by peripheral fishes in the school, evasive action depends mainly on the speed of transmission of warning signals to all members of the school.

### 3.2.3.3 "Alarm Reaction"

Although predator avoidance by schooling fishes often seems to be mediated by visual cues, both from attacking predators and from disturbed members of the school, one well-known escape response is initiated by chemical cues. This is the "alarm reaction" ("Schreckreaktion") described first for the European minnow *Phoxinus phoxinus* by von Frisch (1938, 1942). In this reaction fish show an intense, species-typical flight response to substances released from injured conspecifics. The effective material ("Schreckstoff") is produced in specialized epidermal club cells, and is released into the water only when those cells are damaged (Pfeiffer, 1962). In aquarium-held *P. phoxinus* the first fish to detect the alarm substance dart away from it, swim about in an agitated manner near the bottom, where they are quickly joined by others in the same tank, and all fish then hide under any available cover. Transmission of this reaction to other individuals can be visually mediated, since fish in an aquarium next to one treated with Schreckstoff will respond to the sight of fish performing the Schreckreaktion by showing the same alarm behaviour themselves (Verheijen, 1956; Schutz, 1956).

The alarm response has been documented for many species, all within the superorder Ostariophysi; it occurs primarily but not exclusively among schooling fishes (Pfeiffer, 1962, 1963, 1977). The effective substance in different species is evidently very similar in chemical structure (Pfeiffer, 1967), and this explains the common occurrence of interspecific alarm reactions (Schutz, 1956).

The presumed primary function of the Schreckreaktion is to warn other fishes of the imminent danger of attack. However, there is little direct experimental evidence of its effectiveness as an anti-predator mechanism. Indirect evidence includes the nature of species within the Ostariophysi that lack the reaction. Among these are blind cave fishes (*Anoptichthys*), piranhas (*Serrasalmus*), armoured catfishes (Loricariidae), banjo catfishes (Aspredinidae), and knife-eels (Gymnotidae). All of these fishes have efficient means other than schooling to avoid predation; they are either nocturnal, cave-dwelling, armoured, electric, or strongly predaceous (Pfeiffer, 1977).

### 3.2.3.4 Schooling as Shelter-Seeking

Williams (1964) argued that schooling behaviour can be seen as the combined result of the actions of individual fish, each one using others in the group as a form of shelter from predators. The concept of natural selection favouring such apparently "selfish" behaviour was further developed by Hamilton (1971) and Vine (1971). It follows from this argument that if fish in a group do show a centripetal tendency to move toward the centre, and if predators strike first at the most isolated group members, the net effect should be dense, closely packed schools, with peripheral members continually trying to shift inward. Although predators do tend to strike at isolated individuals, there is little evidence that peripheral members of fast-moving, polarized schools regularly shift inward. On the other hand, young catfish (*Ictalurus melas*) live in tightly packed, slow-moving schools in shallow water, and there is a continual pushing inward of fish from the outside (Bowen, 1931). Filmed records of the behaviour of fish in large, free-living schools under predatory attack may be needed to provide convincing evidence on this point.

An extension of Williams' argument is that the incidence of schooling among small, soft-bodied, vulnerable fishes should bear an inverse relation to the presence of effective alternative shelter-sites. Although schooling is common among open-water, pelagic fishes, it is by no means rare in shallow-water species that live close to an abundance of substrate shelters (Breder, 1951; Shaw, 1978). In such situations the opportunity for shelter among fellow group members is not likely to be the only selective advantage of schooling.

### 3.2.3.5 Inhibition of Attack

Some authors have suggested that when many fish are tightly aggregated, the group may from a distance resemble a single larger object or animal that inhibits or deters a predator's attack (Springer, 1957; Breder, 1967). An example is the small marine catfish *Plotosus anguillaris*. When the juveniles are alarmed they clump into a spherical mass, in contact with each other and all orienting head-outward. This clump is said to resemble a sea urchin, and thereby deter predators from attacking (Knipper, 1953). The same species was studied by Magnus (1967b) who did not support the sea urchin-resemblance hypothesis, but did agree that the schools of young *P. anguillaris* looked like large, moving objects. Neither author provided evidence on predator responses to the schools.

On the other hand, many observers have recorded the apparent reluctance of large piscivores to attack a cohesive, massed school of smaller fish, while quickly snapping up individuals that stray from the school (e.g., Hobson, 1968a; Radakov, 1973). Although a satisfactory explanation of the reluctance to attack large, well-organized schools is not available, it seems unnecessary to postulate that predators perceive them as larger animals. Possibly some predators learn by experience that direct attacks on a polarized school are often unsuccessful, whereas individuals that behave differently or become separated from the others are easier prey.

### 3.2.3.6 "Confusion Effect"

The most commonly proposed mechanism by which schooling protects fish against predation is the so-called "confusion effect". When a visually hunting predator attacks a group of prey that are of a size that must be taken singly rather than several at once, it will choose one among many targets. If these are all moving along independent, erratic paths (due to "flash expansion", streaming to and fro, or other forms of non-polarized locomotor behaviour), the predator may hesitate before attacking. If the hesitation is prolonged, the prey may reform into an organized school and move away from the attack zone. This is an attractive hypothesis. It is easy to see that the sudden, simultaneous performance of erratic, protean, escape movements by many closely grouped animals can present a confusing stimulus situation. The predator encounters many stimuli that elicit conflicting orienting responses; hence the result is "confusion" (Humphries and Driver, 1970). Any delay in attack, however brief, serves the interests of the prey fish.

Support for the hypothesis also comes from the direct observation of predators attacking schools under natural conditions (Eibl-Eibesfeldt, 1962; Hobson, 1968a;

Potts, 1970; Nursall, 1973). Often the attackers show the effects of the multiplicity of conflicting stimuli by hesitating, withdrawing, or missing when trying to capture a single fish. In summary, "...efficient hunting requires most predators to direct their strikes at individual prey, and...such attackers experience difficulty when confronted by the many alternate targets presented by the school" (Hobson, 1968a, p. 87).

And yet, one must use care in generalizing here. There is ample evidence from field observers that predators are in fact often stimulated to attack schooled prey at the very point where some of the fish are disoriented and moving in an erratic manner (Eibl-Eibesfeldt, 1962; Hobson, 1968a; Radakov, 1973). Since these points of disruption may be caused by fish being injured or disoriented by previous attacks, the predators may well increase their chances of success by attacking there. In general, it may be that the "confusion effect" is an effective anti-predator mechanism only in situations where predators hunt visually, take one prey at a time, and where the prey remain vigorous and active in spite of following erratic, unpredictable swimming paths. Rigorous assessment of the effectiveness of this mechanism will only come from controlled experiments, and very few have been done.

In one experiment Radakov (1973) found that cod (*Gadus*) were able to catch solitary, underyearling pollock (*Pollachius*) in a tank in the mean time of 26 s (n = 23), but did not catch the first fish from a school of pollock until 135 s had elapsed (n = 18). The explanation for this difference seemed to be that a cod pursuing several prey frequently changed the target of its chase before finally attacking one. With a single prey the chase was direct and persistent.

In a more elaborate experiment Neill and Cullen (1974) measured the effect of prey group size on the hunting success of four different predator species. Predators and prey tested together were: cuttlefish (*Sepia officinalis*) and young mullett (*Mugil* sp.), squid (*Loligo vulgaris*) and *Atherina* spp., pike (*Esox lucius*) and young cyprinids, and perch (*Perca fluviatilis*) and female poecilids (*Poecilia vivipara*). Results were presented as the ratio of captures to contacts (see Neill and Cullen for definition of contact, which varied among the predators). For all four predators the capture/contact ratio declined as the number of prey increased (Fig. 3.8). That is, predator attack success per encounter decreased with increasing prey density. The explanation for these results depended on the hunting technique of the predator. Perch actively chased their prey and as school size increased, they more frequently switched targets and with each switch reverted to an earlier stage in the hunting sequence. This led to lower success rate with larger prey numbers, and appears to be a measure of the "confusion effect". The other three predator species were considered to be ambush predators, and increasing prey numbers reduced their hunting success by causing frequent performance of acts not related directly to hunting, hence termed "irrelevant" and "avoidance" (Neill and Cullen, 1974). Interruptions in hunting caused by these acts were responsible for lower capture rate with increasing prey numbers.

The value of this experimental technique lies in the fact that detailed records were kept of all behaviour of the predators after exposure to prey. This allowed Neill and Cullen to quantify and hence compare the behaviours that interfered with hunting.

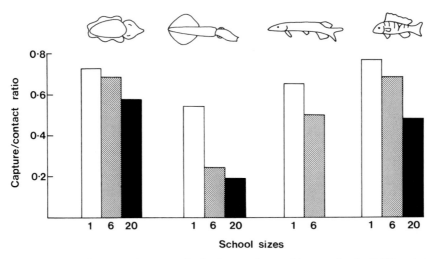

**Fig. 3.8.** Change in capture/contact ratios by four predators with prey schools of different sizes. (After Neill and Cullen, 1974)

## 3.2.4 Aggressive Defence

Many fishes have physical or chemical weapons that are said to be used in defence against attacking predators. If such weapons induce a predator to stop its attack, or to release a captured prey, then their effectiveness as anti-predator mechanisms is confirmed. The problem is to obtain unequivocal evidence that these natural weapons are effective in countering predatory attacks. This is especially difficult in the case of chemical weapons.

Anyone who has handled live, spiny-rayed fishes knows that the prick of a strong, sharp spine can be painful. When a venom gland is associated with the spine, a stabbing wound by such a device may have serious effects. The graphic description by Steinitz (1959) of the effects on him of a wound by the spines of a lionfish, *Pterois volitans*, is evidence of the extreme potency of such weapons.

On a less dramatic scale, most spiny-rayed fishes expand their fins fully when alarmed. Often the fish also rolls sideways with its dorsal spines directed towards the alarming stimulus. If the latter is a predator that continues to advance, the prey fish may attack, striking at the offender with its spines. This form of aggressive defence has been described for scorpionfishes (*Scorpaena*) and gurnards (*Dactylopterus*) (Hinton, 1962; Breder, 1963). The South American doradid catfish *Hassar orestis* when placed in an aquarium containing a single piranha *Serrasalmus nattereri* will try to find shelter on the substrate. If none is available it positions itself directly below the piranha with its large dorsal and pectoral spines fully erect (Fig. 3.9). If the piranha tries to catch the smaller fish it is pricked in the throat or belly by the dorsal spine, and eventually gives up attacking. *H. orestis* that remained below the piranha survived much longer than those that swam about the tank (Markl, 1968). The frequency of such behaviour in nature is unknown; presumably substrate shelters will be used when available, even by fishes with effective natural weapons.

58

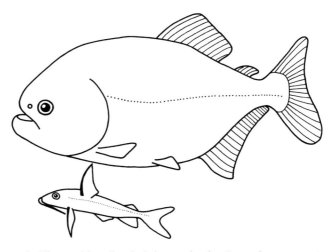

**Fig. 3.9.** The catfish *Hassar orestis* holding position directly below a piranha *Serrasalmus nattereri.* (After Markl, 1968)

Porcupinefishes (Diodontidae) and puffers (Tetraodontidae) can inflate themselves when threatened by swallowing water so that their body size increases greatly and the many sharp spines on the body stand erect. This formidable appearance and set of weapons should be a deterrent to attack, but again there is little direct supporting evidence (Marshall, 1965).

One experimental study showed that spiny-rayed fish can use their spines to derive protection against predators. Hoogland et al. (1956) found that pike (*Esox lucius*) and perch (*Perca fluviatilis*) learned by experience to stop attacking sticklebacks. They avoided the species with fewer but larger spines (*Gasterosteus aculeatus*) more consistently than the species with smaller but more numerous spines (*Pungitius pungitius*). The experience causing avoidance was apparently the difficulty the predators had in swallowing the prey. They often captured and spit out a stickleback quickly. Such rejection behaviour was more frequent for *Gasterosteus* than *Pungitius*. Pike often coughed or made violent head movements, especially after capturing a *Gasterosteus*. The latter species is able to lock its spines in position when fully erect (Hoogland, 1951), and the large spines frequently jammed in the predator's mouth (Fig. 3.10). Although hungry pike eventually ate even such awkward prey, the effect of the spines on predation was clearly shown in an experiment where a pike was presented with equal numbers of the two stickleback species and *Phoxinus phoxinus* (a cyprinid with no fin spines). Although all prey fish were eventually eaten, the cyprinids were taken first, and *Gasterosteus* survived longer than *Pungitius* (Fig. 3.11). In nature any spine-induced delay in predation would give sticklebacks an opportunity to escape.

Several species of the poison-fang blenny *Meiacanthus* have one long sharp canine tooth at the rear of each lower jaw. The teeth have venom glands at the base, and when a *Meiacanthus* is taken into the mouth of a large predator it inflicts a bite that usually causes the predator to reject it unharmed (Springer and Smith-Vaniz, 1972). A typical predator reaction is "... violent quivering of the head with

59

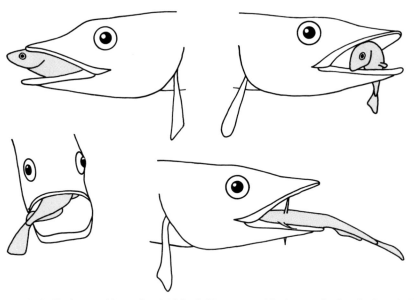

**Fig. 3.10.** Various positions of a stickleback (*Gasterosteus*) in the mouth of a pike (*Esox*). (From a film, after Hoogland et al., 1957)

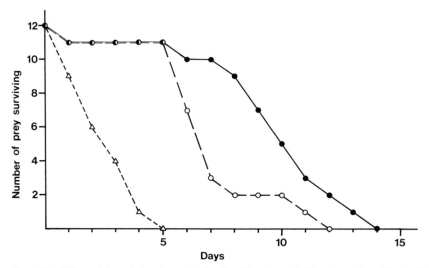

**Fig. 3.11.** Differential predation by a single pike (*Esox*) on 12 *Phoxinus (dotted line)*, 12 *Pungitius (broken line)* and 12 *Gasterosteus (solid line)* over 14 days. (After Hoogland et al., 1957)

distension of the jaws and opercula. The fish frequently remained in this distended posture for several seconds until the *M. atrodorsalis* emerged..." (Losey, 1972, p. 135). When some *M. atrodorsalis* had their fangs removed they were eaten readily by groupers (*Epinephalus merra*), whereas the same predators rejected intact blennies (Losey, 1972). The diet of the poison-fang blennies is plankton and small benthic invertebrates. The fangs are not used for feeding; they seem to function solely to induce rejection after capture by a predator.

60

There is no way of knowing, at present, whether the rejection of prey that attack with both physical (teeth and spines) and chemical weapons is caused by pain from the stabbing wound, from the venom, or both. The almost instantaneous response of *Epinephalus* to attack inside the mouth by *Meiacanthus* suggests that puncture of tissue is alone responsible. The venom usually takes a longer time to induce its major effect, at least in humans.

## 3.3 General Comments on Anti-Predator Behaviour

Although it has been convenient to discuss anti-predator behaviour as a series of discrete defensive mechanisms, this is not meant to imply that each species uses only one such mechanism. In fact the strategy used to avoid being eaten can vary in two quite different ways. First, a fish may switch from one tactic to another if a persistent predator continues to attack. For example a stickleback (*Gasterosteus*) may be well-hidden in dense vegetation until a persistent hunting perch (*Perca*) discovers it and attacks. The stickleback then flees along an erratic path, with the predator chasing. Eventually the fleeing fish is cornered, and when snapped up by the perch, erects all its spines. If the perch is not large enough to swallow the prey readily, it may be stabbed inside the mouth by the spines, and quickly spit the prey out. After repeated attempts to swallow the stickleback, the perch leaves it alone and resumes searching for other prey.

Second, most fishes have active periods, when they are feeding and possibly breeding, and inactive periods, when they rest. The anti-predator behaviours most suitable for these periods are likely to be quite different. To take *Gasterosteus* again, the overnight resting period is likely to be spent in hiding close to the substrate. Cryptic colouration provides it with camouflage. By day it feeds on micro-organisms and plankton. Some feeding populations are relatively benthic and solitary; when a predator approaches they are likely to dash for cover and hide. Others are more pelagic; they feed in schools and use one or more of the schooling mechanisms to avoid capture when attacked (Larsen, 1976). For many species, regardless of whether they are diurnally or nocturnally active, the twilight changeover period between activity and resting is the most dangerous time of the diel cycle (Hobson, 1972, 1974). They are then either leaving shelter and moving to a feeding area, or vice versa. In either case, their vulnerability is increased, since neither their nocturnal nor their diurnal anti-predator mechanisms is likely to be fully effective.

There is a need for more experimental data bearing directly on the relative survival values of different defence mechanisms. It seems to be remarkably difficult, for example, to obtain unequivocal evidence on the protective value of schooling. A series of studies by Seghers showed that useful insights into the selective value of certain behaviour patterns can come from a combination of detailed field observations and laboratory experiments using natural predators. Seghers (1974a) found that allopatric populations of the guppy (*Poecilia reticulata*) in Trinidad showed different intensities of schooling. Fish from areas where predators were numerous schooled more closely than those from streams with few predators. Breeding experiments with predator-naive fish showed that these differences were genetically determined. When guppies from different populations were simultaneously exposed to the piscivorous pike-cichlid *Crenicichla alta*, losses to the predator were consistently highest among fish from the less intensely schooling stocks. A precise explanation of the advantage of schooling was not clear, however, since other work had shown that these guppy populations also differed in other features that could be related to defence. These included reaction distance to predators, threshold for alarm, and microhabitat selection (Seghers, 1974a,b).

Chapter 4

# Selection and Preparation of Spawning Site

Reproductive behaviour in fishes, as in other animals, has received a great deal of attention from ethologists. The literature is so large, and the subject so important (not least to the fish themselves), that I have devoted three chapters to it: this one on selection and preparation of a site for spawning, Chapter 5 on courtship and the spawning act itself, and Chapter 6 on parental care of the young.

There is great diversity in the extent to which fishes prepare the site where they spawn. Those with buoyant, pelagic eggs generally spawn in open water, well above the substrate, and have no specific site to prepare. Even among species with demersal eggs, many do not modify the spawning location; gametes are released synchronously and the fertilized eggs settle onto the substrate. However, many fishes do prepare a site, either by clearing off a firm substrate and making it suitable for the attachment of eggs, or by modifying it so as to enhance protection of the developing eggs.

## 4.1 Spawning Surface Cleaned

Some fishes spawn on a firm, stationary surface such as a stone, leaf, or rocky patch on a coral reef. Before breeding, any algae, silt or debris covering the site is removed. Since most substrate-spawners lay adhesive eggs, this preliminary cleaning presumably increases the likelihood of the eggs remaining firmly attached during their development. The most common motor patterns used for cleaning are two mouth-contact behaviours. One is a rapidly repeated biting and scraping at the surface, with the fish occasionally moving away and spitting out accumulated material. The second is a slower, more deliberate action in which the fish contacts the substrate with its mouth open, vigorously snaps the mouth shut and bounces a short distance away. This may be repeated many times. Both behaviours were called "nipping" by Baerends and Baerends-van-Roon (1950), although it may be useful to distinguish between them by referring to the bouts of rapid biting as "nibbling", and the slower, snapping movements as "nipping" (Polder, 1971). Figure 4.1 shows two fishes nipping at potential spawning sites. Other motor patterns commonly used in cleaning are sweeping movements of the fins and picking up large particles or whole organisms in the mouth and carrying them away.

Somewhat more specialized techniques are used by some damselfishes (Pomacentridae). For example, a male and female *Pomacentrus leucoris* in an aquarium picked up sand in the mouth and spat it out against a vertical rock face,

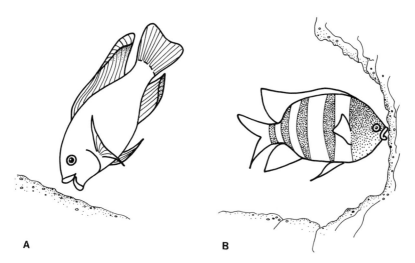

**Fig. 4.1.** Cleaning the potential spawning site by mouth-nipping at the substrate. **A** *Cichlasoma biocellatum*. (After Baerends and Baerends-van Roon, 1950). **B** *Abudefduf saxatilis*. (After Fishelson, 1970)

vigorously fanned the area with their fins, and then picked remaining sand grains off the rock with their mouths. After several hours of this "sand blowing", fanning and picking, the pair spawned on the cleaned area (Breder and Coates, 1933).

The garibaldi (*Hypsypops rubicunda*), a pomacentrid of the rocky intertidal zone of southern California, prepares a spawning site consisting of a 15 to 40 cm diameter clump of red algae. An adult male maintains a territory at the site throughout the year. During the breeding season he actively crops the algal patch to a height of 2 to 4 cm and clears all loose materials from a narrow zone surrounding the algal nest, in which the eggs are laid. Outside the breeding season the male ceases the cropping and clearing activities, and the area soon becomes overgrown with other organisms (Limbaugh, 1964; Clarke, 1970).

Table 4.1 summarizes the spawning site cleaning behaviour of several pomacentrids, to illustrate the variety of methods used. In most cases the male alone does the cleaning, but in species where a male–female pair guards the brood together, both sexes prepare the site.

## 4.2 Spawning Site Excavated

Some fishes prepare a spawning site by excavating a pit in the substrate within which the eggs are laid. Four types of excavation can be distinguished, based on the kind of protection the newly fertilized eggs receive: that in which the eggs are buried after spawning and the parents soon leave; that in which the eggs remain exposed, but are guarded by one or both parents; that made inside or underneath a natural shelter; and that from which the eggs are carried away by one parent immediately after spawning.

**Table 4.1.** Spawning location, sex of cleaner, and methods used to clean spawning sites of some Pomacentridae

| Species | Spawning location | Sex of cleaner and cleaning methods used | Authority |
|---|---|---|---|
| *Abudefduf saxatilis* | Concavities in vertical walls and horizontal surfaces of reefs | ♂ Nibbles with teeth, carries sand and pebbles in mouth, fans | Fishelson, 1970 |
| *Chromis multilineata* | Stone, base of gorgonian or seaweed | ♂ Carries away hermit crabs, bites off *Diadema* spine tips, fans | Albrecht, 1969 |
| *C. caeruleus* | Green algae clumps on reef crest | ♂ Nips and skims over algae | Sale, 1971a |
| *Pomacentrus leucoris* | Vertical rock face | ♂ and ♀ spit sand at site, fan and pick sand off | Breder and Coates, 1933 |
| *Dascyllus aruanus* | Base of branching coral colony | ♂ Nips algae and skims | Sale, 1970 |
| *D. marginatus* | Base of branching coral colony | ♂ Nibbles algae | Holzberg, 1973 |
| *Hypsypops rubicunda* | Red algae patch | ♂ Crops algae patch, clears belt around algae by biting, carrying away large objects | Limbaugh, 1964; Clarke, 1970 |
| *Amphiprion* spp. | Stone at base of sea anemone | ♂ and ♀ remove particles by nibbling | Wickler, 1967 |

## 4.2.1 Eggs Buried

These fishes lay a batch of eggs in a pit, and then cover them with loosened substrate materials. Often the egg-burying activities also serve to excavate another spawning pit, close to the first, in which more eggs are laid and buried. The result of a series of these digging, spawning and burying activities is a nest containing one or more separate batches of fertilized eggs buried under protective substrate materials. Normal egg development depends on the ability of the adults to select a suitable site, and on local environmental conditions during incubation and early growth. This type of spawning site selection and preparation is well illustrated by three groups of fishes that breed in freshwater: lampreys (Petromyzontidae), Salmonidae and some Cyprinidae.

### *4.2.1.1 Petromyzontidae*

All lamprey species, whether anadromous or strictly freshwater, spawn in pits excavated in shallow, freshwater streams. The parasitic Atlantic sea lamprey (*Petromyzon marinus*) is the most intensively studied species, because of the devastating effect it has had on native fishes of the Great Lakes of North America, after gaining access to the lakes via shipping canals. A detailed study of *P. marinus* showed that a suitable sand and gravel stream bottom and adequate water flow are the most critical requirements for successful spawning (Applegate, 1950). Substrate particle size is most important. Silt and sand without gravel are too unstable for a durable nest, and as anchor for spawning lampreys, while large rocks are too heavy or too deeply imbedded to be moved.

Both particle size and degree of illumination influence nest site selection in the European lamprey, *P. fluviatilis*. In aquaria with uniform sand and stone substrate, but with one half shaded and the other half receiving weak overhead light, nests were always dug in the shaded half. When the substrate was arranged so that different areas had different sized stones, or plain sand, the lampreys dug in the sand, even when it was unshaded (Hagelin and Steffner, 1958). When filming of spawning required nests to be built at the front of the aquarium, lampreys were induced to dig there by increasing the proportion of fine compared to coarse gravel at that location (Hagelin, 1959).

Two main excavating techniques are used by all species that have been observed (Hardisty and Potter, 1971). In one, the lamprey attaches its suctorial oral disc to a single stone and swims away with it. Occasionally two lampreys fasten their discs to a single large stone and move it together (Dendy and Scott, 1953). The second method involves the lamprey attaching itself by the oral disc to a large, well-imbedded stone at the upstream edge of the nest area and vigorously undulating its body and tail. Often the animal turns over on one side and the flapping movements are then mainly in the vertical plane (Lohnisky, 1966). This behaviour loosens stones, sand and fine particles in the bottom of the nest depression, and the water current carries the lightest material downstream. As digging continues, an excavation is gradually formed that is oval to circular in shape, has gravel and large stones around the upper edge and sides, and finer materials on the downstream side. The dimensions of the nest depend on the character of the substrate materials, the speed and depth of water flowing over the area, and the size and number of lampreys involved in the excavation (Hardisty and Potter, 1971). Two views of a typical nest of *Lampetra planeri* are shown in Figure 4.2.

Both sexes contribute to nest construction in all species studied (Scott and Crossman, 1973), although frequently males begin the excavating and females join in the digging as the breeding season progresses (Hagelin and Steffner, 1958). Early in the season male *P. marinus* often dig several incomplete "trial" nests, and abandon them before settling down to complete a nest at one location (Applegate, 1950).

The number of lampreys excavating at a single site varies widely with the species. Whereas *P. marinus* nests are mostly made by a single pair of adults (Applegate, 1950), up to 20 or more adults have been observed excavating a single nest of several smaller species (Dendy and Scott, 1953; Hardisty and Potter, 1971). An extreme case of communal excavating was observed in the chestnut lamprey, *Ichthyomyzon castaneus*, by Case (1970). About 50 individuals excavated two adjacent but separate nests, which gradually fused into a single one as each was enlarged. As spawning proceeded some lampreys moved stones from the upstream edge of the nest back into the excavation, presumably covering eggs which had been released there. In this way the nest slowly expanded in size so that 24 h after fusion of the two original nests, the entire structure occupied an oval-shaped area about 6 m long and 1 m wide. Such expansion in size of a single, communal nest is not known for other lamprey species (Case, 1970).

After spawning, all lampreys anchor themselves to stones at the head of the nest and with vigorous body undulations loosen up more sand and gravel which settles downstream, covering and imbedding the eggs where they have settled among the

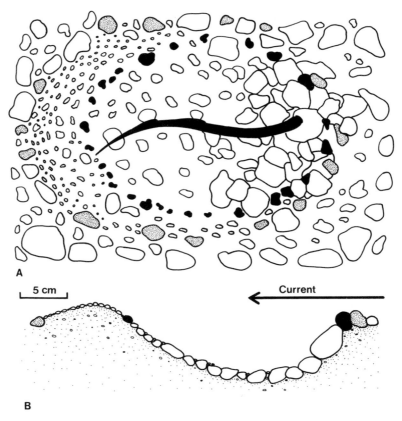

A

L 5 cm ₁                                          ⟵ Current

B

**Fig. 4.2. A** *Lampetra planeri* in its nest, seen from above, **B** side view of same nest, showing distribution of substrate materials. (After Lohnisky, 1966)

stones. Often larger stones are moved individually to the downstream slope of the nest. The end result is a stable structure, incorporated into the substrate, and within which the fertilized eggs are protected from predation and the effects of fluctuations in stream flow during development.

### 4.2.1.2 Salmonidae

Most species of salmon, trout and char spawn in the gravel substrate of relatively fast-flowing, shallow rivers and streams. Some populations spawn along lake shores, usually near stream inlets or subterranean springs (Scott and Crossman, 1973). The adult female selects a site and excavates a depression in which spawning occurs. After one batch of eggs is laid, she resumes excavating immediately upstream from or adjacent to the first site. This creates a new depression where more eggs are deposited and the displaced sand and gravel bury the eggs previously laid. When all her eggs are laid the spawners depart, leaving behind a nest or *redd*, with several batches of fertilized eggs buried in it.

Successful salmonid redds depend on two main criteria: (1) sufficient water flow through the gravel to provide oxygen for the eggs and to remove metabolic wastes,

**Table 4.2.** Number of female char (*Salvelinus alpinus*) that began nest-digging in each of seven substrate zones in a stream tank (Fabricius and Gustafson, 1954)

| Bottom materials | Diameter of materials (mm) | No. of ♀♀ |
|---|---|---|
| Fine sand, sparsely planted *Isoetes* | < 0.6 | 0 |
| Fine sand, no plants | < 0.6 | 0 |
| Coarse sand | 0.6–5 | 1 |
| Fine gravel | 10–15 | 1 |
| Coarse gravel | 20–60 | 9 |
| Cobblestones | 100–250 | 4 |
| Patches of sand and gravel between cobblestones | Patch diameter < 100 mm | 2 |

and (2) gravel of such a depth and size composition that the buried eggs are protected against scouring action of floods, desiccation during periods of low water, sudden vibrations, and direct sunlight (Foerster, 1968).

Little experimental work has been done on the cues salmonids use to select digging sites. Fabricius and Gustafson (1954) studied spawning of Arctic char (*Salvelinus alpinus*) in a large artificial stream tank the bottom of which was divided into seven 1 m² zones, each with distinctive bottom materials. Of 17 female char that spawned in the tank, initial digging by most of them occurred in zones with coarse gravel or cobblestone substrate (Table 4.2). Their behaviour indicated that visual cues were predominant in digging-site selection. They swam close to the bottom, frequently tilting the head and directing the eyes downward. When a large glass plate was placed on a substrate consisting either of areas of fine sand and coarse gravel, or of flat stones and coarse gravel, female char appeared to inspect the substrate through the glass, and when digging was performed on the glass it was always over the gravel area (Fabricius and Gustafson, 1954).

The range in gravel size used for redds by various species depends, in large measure, on the size of the spawning fish. Smaller fish construct smaller redds in finer gravel than do larger fish (Briggs, 1953; Hartman, 1969). Likewise larger fish dig successful redds in areas with higher water velocity than do smaller fish (Hartman, 1970). These relationships no doubt stem from the greater physical strength and hydrodynamic stability of the larger fish.

Accelerating water flow seems to be a strong stimulus for nest site selection. In large, flowing-water tanks, rainbow trout (*Salmo gairdneri*) and chum salmon (*Oncorhynchus keta*) excavated nests in areas of upwelling, accelerating current just upstream from gravel or rock mounds that reduced water depth and changed the pattern of flow (Tautz and Groot, 1975).

*Salmonid Nest-Digging Behaviour.* A bout of digging begins with the female facing upstream, then turning over on one side and vigorously flexing the tail and posterior part of the body several times in rapid succession so that the caudal fin beats up and down against the gravel. The downstroke of each flexure loosens the gravel by forcing water down against it, and the upstroke moves loose gravel out of the redd by suction (Jones and King, 1950). Between digging bouts the fish may settle into the bottom of the excavation and press her extended anal fin down

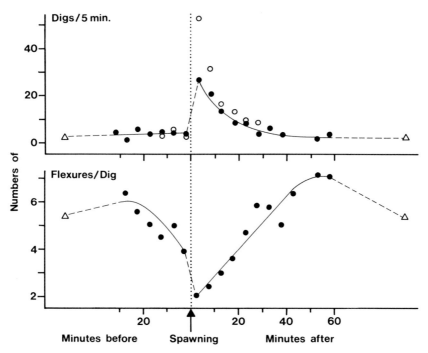

**Fig. 4.3.** Redd-digging behaviour of female sockeye (*solid circles*) and kokanee salmon (*open circles*) shortly before and after spawning act. *Triangles*, mean values for all female sockeye observed. Lines fitted to sockeye points by eye. (After McCart, 1969)

among the stones. This "feeling" or "probing" act is common to all stream-spawning salmonids (Needham and Taft, 1934; Smith, 1941; McCart, 1969; Hartman, 1970). It appears to be a method of testing the suitability of the redd for egg deposition, and increases in frequency just before spawning (Jones, 1959).

Many observers have noticed that the digging rate by female salmonids increases after a bout of spawning, as she covers the eggs with gravel. Supporting quantitative data were provided for sockeye salmon (*O. nerka*) by McCart (1969). During the 30 min before a spawning act, the rate of digging changed very little, but the mean number of tail flexures per digging bout declined "...as females seem to shift their emphasis from simple excavation to a more careful finishing of the nest" (McCart, 1969, p. 43; Fig. 4.3). Immediately after spawning the frequency of digging bouts increased sharply, both for anadromous sockeye and for kokanee, the smaller, landlocked form of the species. The number of tail flexures per digging bout was low at first, but gradually rose to a level above the pre-spawning peak. The effect of this post-spawning behaviour was that the female maintained a high digging rate during the critical period when the eggs were first being covered by gravel. Both digging rate and tail flexures per dig gradually returned to pre-spawning levels by 30 to 40 minutes after spawning (Fig. 4.3).

Similar changes in redd-digging by female rainbow trout (*S. gairdneri*) immediately after spawning were recorded by Hartman (1970). Digging rate increased and then declined to pre-spawning levels; the number of body flexures per

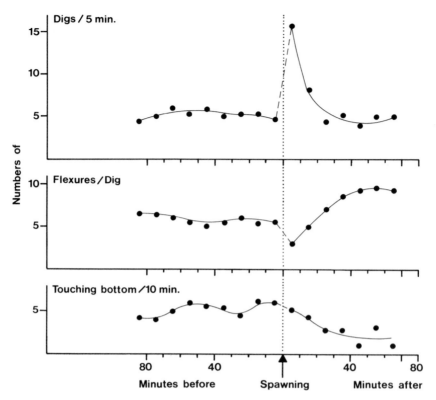

**Fig. 4.4.** Redd-digging behaviour of female *Salmo gairdneri* shortly before and after the spawning act. Based on 13 fish. (After Hartman, 1970)

digging bout decreased and then rose sharply and remained above pre-spawning levels until at least 70 min after spawning (Fig. 4.4). Hartman also found that the frequency of contacting the substrate with anal or pelvic fins increased gradually before spawning, and then declined (Fig. 4.4). He suggested that the female fish obtained tactile information about the condition of the nest by probing into the gravel with her anal fin. Before spawning this information could be used to indicate when the site was suitable for receiving eggs; after spawning it could indicate areas that required more covering with gravel. Fish were often seen digging upstream from the precise area of probing.

Despite these demonstrated quantitative differences between post- and pre-spawning digging rates, the motor patterns are virtually identical in both situations for species of *Oncorhynchus* and *Salmo*. This is not the case, however, for the genus *Salvelinus*. Pre-spawning digging is typical in form, but when covering up newly spawned eggs, the female remains upright over the redd with the head held higher off the gravel than the tail. She then undulates the tail and body in a slow, somewhat rhythmical manner while moving about over the redd. During this activity, the pelvic, anal, and caudal fins sweep any eggs that are on the surface of the gravel down among the stones, and also gently push stones in from the sides of the redd to cover the eggs. After a bout of undulating, the female resumes typical

70

salmonid excavating behaviour in preparation for another spawning bout. Post-spawning undulating as described has been observed in brook trout, *Salvelinus fontinalis* (Needham, 1961), Dolly Varden, *S. malma* (Needham and Vaughan, 1952), and the Arctic char, *S. alpinus* (Fabricius and Gustafson, 1954).

Occasional redd-digging by male salmonids has been reported by several workers. McCart (1969) provided detailed descriptions for *O. nerka*. The male motor patterns appeared to be identical to those of females, but were less frequent and vigorous. Male digging was performed by a male closely associated with a female on a redd, and occurred either when another male approached, or when a spawning bout was interrupted by another male (both situations usually eliciting aggression). In either case male digging appeared out of context, and was associated with possible conflict between the motivations to attack and withdraw, or to attack and continue behaving sexually. In any event, it resembled a "displacement activity", and was seen often enough for McCart to conclude that it was a natural event, not an aberration. On the other hand, there was no indication that excavations dug by males were ever formed into complete redds, or that spawning occurred in them (McCart, 1969).

### 4.2.1.3 Cyprinidae

Some North American cyprinids breed in the same general manner as lampreys and salmonids. That is, they dig an excavation in the gravel substrate of flowing streams, spawn in it, and cover the eggs with gravel. An example is the river chub, *Nocomis micropogon*. A mature male begins the next excavation by picking up small stones or sand in his mouth and dropping them nearby. He may push against larger stones with his opened mouth. When a pit has been excavated the male gradually reduces the frequency of carrying stones away, and begins to carry stones back into the pit. The same motor patterns are used for pit-filling as for the earlier excavating. As this work continues a dome-shaped nest is gradually constructed that, when completed, is larger in diameter than the original pit, and may be 15 to 20 cm above the level of the surrounding substrate (Raney, 1940a; Reighard, 1943). The main function of all this effort seems to be to prepare a nest of accumulated stones relatively free of sand and other small particles. During nest-building one or more mature females may approach the male, who quickly excavates a small trench on top of the gravel pile. This serves as a spawning site, and if eggs are laid in the trench the male quickly fills it in with more stones. Occasionally, when the male is away from the nest, females may excavate small depressions on the nest surface, but spawning generally occurs in trenches made by the male. In this way the nest gradually increases in size and when completed it contains several batches of fertilized eggs in small pockets. Other cyprinids (e.g., *Notropis cornutus* and *N. rubellus*) may spawn on the large nests of *N. micropogon*, but the male continues his stone-carrying behaviour even when other fishes are swarming over his nest (Raney, 1940b).

The creek chub, *Semotilus atromaculatus*, spawns in shallow, gravelly streams where each adult male establishes a territory. He excavates a shallow pit by picking up stones in his mouth and dropping them slightly upstream. Then one or more females approach, spawning occurs in the pit, and the male buries the eggs with more gravel carried from the area immediately downstream. This sequence is repeated as the male spawns with many different females, and the result is a steadily enlarging nest, continually being extended downstream. When complete it has the form of a long, low, narrow ridge, which may be 25–30 cm wide, 6–8 cm high and up to 5 or 6 m in total length (Reighard, 1910; Raney, 1940a).

Finally the male cutlips minnow, *Exoglossum maxillingua*, selects a nest site, often near a submerged log, excavates a shallow depression and then carries stones there from a wide surrounding area (Van Duzer, 1939). The stones appear to be carefully selected; often several are picked up and rejected before one is carried to the site. In any nest the stones are similar in size and vary with the size of the nesting male, presumably as a function of his mouth size. Spawning occurs on the upstream slope of the nest and the male carries stones to that spot from the nest periphery to bury the eggs.

*General Comments.*The most important requirements for a successful spawning site among fishes considered in this section are gravel of a certain size range and flowing water. The gravel must be small enough to be moved, and yet large enough to remain stable after spawning is complete, despite fluctuations in water flow. Placement of eggs deep within the nest or redd provides protection from predators and from the scouring action of floods. The nest-digging activities of the adult fish are essential to remove sand and other fine material from the gravel bed, so that water continues to percolate among the eggs. The key to successful incubation and hatching of the eggs is this percolation of the surrounding water, providing oxygen and removing metabolic wastes (Wickett, 1954; Pollard, 1955). Even if the upper levels of a redd become clogged with silt, the young fish can develop successfully if fresh water continues to reach them by relatively slow movement through deeper levels.

## 4.2.2 Eggs Exposed

Some fishes spawn in relatively open, exposed areas, where they excavate a pit in the substrate, spawn in it and then guard the young until they disperse. The excavation serves both as a spawning site and a shelter for the protection of the eggs and newly-hatched young. The North American sunfishes (Centrarchidae) illustrate this system well.

### 4.2.2.1 Centrarchidae

Eleven species are currently recognized in the sunfish genus *Lepomis* (Bailey, 1970). Reproduction is similar throughout the group. In spring or early summer mature males move into shallow water along the margins of ponds, lakes or streams. Each male establishes a territory and in it excavates a shallow nest, where spawning occurs. The male guards the progeny until they scatter. Male *Lepomis* excavate their nests by using the tail or the mouth. The commonest technique is a vigorous, lateral sweeping movement of the tail that directs water downwards, stirring up the substrate. With repeated digging, lighter materials (silt and detritus) are dispersed from the area, and heavier particles are gradually shifted to the nest periphery. The completed *Lepomis* nest is a bowl- or saucer-shaped pit, whose dimensions are

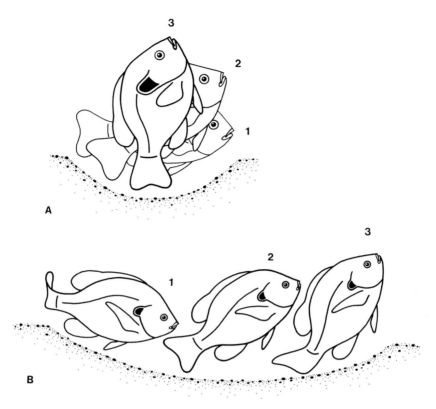

**Fig. 4.5.** Change in orientation and position in the nest of male *Lepomis megalotis* (**A**) and male *L. gibbosus* (**B**) during a nest-digging bout. *Numbers* refer to successive positions of the male during a single bout

determined by the composition of the substrate, the size of the digging fish, and most importantly the extent of horizontal movement by the fish during digging bouts. Observations in the field and in aquaria have shown that the males of some species (e.g., *L. macrochirus, L. megalotis,* and *L. humilis*) shift from horizontal to vertical, head-up orientation, with little horizontal movement through the nest as they dig. In contrast, male *L. gibbosus* and *L. auritus* only pitch to an angle of from 20° to 70° head-up, but they move horizontally through the nest while digging (Fig. 4.5). Nests of the former group tend to be relatively small and bowl-shaped, with clearly defined rims and steep sides; nests of the latter species are usually larger, less regular in outline and with more gradually sloping sides.

Despite this variation in structure of completed nests, male *Lepomis* are not rigidly restricted to occupying and spawning in the nest-type "typical" for that species. In fact, males may occupy, clean out and spawn in nests previously used by another congeneric species (Breder, 1936; Clark and Keenleyside, 1967). A shortage of suitable nesting substrate probably contributes to successive use of a single nest by males of different species.

The other common nest-excavating technique of male *Lepomis* is to grasp leaves, sticks, rooted plants, small stones, etc. in the mouth and by carrying, tugging, or pushing, remove them from the site (Leathers, 1911; Fowler, 1923; Miller, 1963). The frequency of these activities in relation to tail-digging depends on the composition of the substrate at the nest site.

The location of *Lepomis* nests has frequently been described, but nest site selection has seldom been studied experimentally. All species are said to excavate nests on firm substrate, in shallow water close to shore, although few attempts seem to have been made to search for them in deep water. Generally the males dig through loose, soft materials until reaching a firmer substrate of gravel, stones, or thickly matted roots of aquatic plants (Barney and Anson, 1923; Breder, 1936; Ingram and Odum, 1941). Proximity to shore may be an important aspect of nest site selection. Clark and Keenleyside (1967) observed that most *L. gibbosus* nests in two small ponds were dug within the 0 and 50 cm depth contours. In one pond, where a large area less than 50 cm deep extended over 25 m from shore, many nests were built in that area, but all were close to shore, suggesting that proximity of the shoreline was a stronger attraction than open water of the same depth.

In a study of nesting by *L. gibbosus*, Colgan and Ealey (1973) established eight experimental areas in a small lake. Each area was divided into two sections. One of these was designated as "cleared"; all loose woody debris over 25 cm long was removed and scattered in the other section, labelled "cluttered". Frequencies of nests excavated by male *L. gibbosus* in the two sections, and other nesting data, were collected throughout the breeding season. The following results were obtained: (1) the most preferred nest sites had thin silt layers over rock or gravel; the least preferred had deep silt over soft clay (2) near the beginning and end of the breeding season, when total numbers of nests were relatively small, more nests were excavated in cleared than in cluttered sections (3) during the peak of breeding, nest densities were not different in the two sections. The explanation for (2) and (3) seems to be that male *L. gibbosus* prefer to nest in relatively open, clear areas when a choice is available, but when a large breeding population limits the available space they will nest wherever the substrate is suitable. Preference for firm over soft bottom suggests the latter is suboptimal habitat, probably because the eggs and young are more subject to smothering there.

### 4.2.2.2 Cichlidae

Some substrate-brooding cichlids spawn at sites that have been cleaned by the courting male and female. The main cleaning pattern is nipping at the surface, by which all loose particles, silt, and attached algae are removed. The frequency of nipping at the eventual spawning site increases shortly before egg-laying begins, even though the surface appears to have been thoroughly cleaned. This has led to the suggestion that it is not only a cleaning action, but is also a signal by one fish to its partner that it is ready to spawn; in other words, it is courtship behaviour (Baerends and Baerends-van Roon, 1950; Baldaccini, 1973). Data supporting this, and showing that the female chooses the spawning site, come from the following two examples.

The angelfish, *Pterophyllum scalare*, is a South American cichlid that spawns on leaves of aquatic plants. Both sexes clean leaves by nipping. Chien and Salmon (1972) found that leaf-cleaning continued even after all available leaves (plastic in

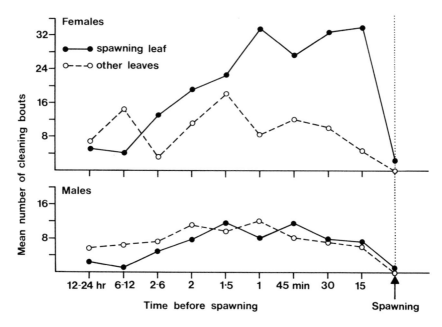

**Fig. 4.6.** Frequency of leaf-cleaning bouts per 15 min by female and male *Pterophyllum scalare* directed at the spawning leaf and at all other leaves during the 24 h before spawning. (Modified after Chien and Salmon, 1972)

this case) had been removed, scrubbed clean with steel wool, and replaced. Up to about 6 h before spawning, females did more bouts of cleaning than males did, and both sexes cleaned several different leaves. Thereafter this difference between sexes increased greatly, and females concentrated their cleaning more and more on the eventual spawning leaf. Males continued cleaning at a lower rate and at a variety of leaves (Fig. 4.6).

*Aequidens paraguayensis* is a South American cichlid that spawns on a loose leaf selected from substrate litter. Both members of a mated pair nip at the surface of leaves with about equal frequency during the two days before spawning, but just before spawning, the frequency of nipping and other leaf contacts by the female increases dramatically (Fig. 4.7). As with female *P. scalare*, the female *A. paraguayensis* indicates her readiness to spawn by repeated mouthing of the site where she is about to lay her eggs (Timms and Keenleyside, 1975).

Male and female *A. paraguayensis* also frequently move leaves before spawning, by grasping the edge in the mouth and pulling, lifting, and turning them. They give the impression of "testing" several spawning substrates before selecting one. When presented with a choice of two plastic leaves differing in area or weight, *A. paraguayensis* pairs consistently spawned on the smaller and the lighter of the two leaves. This was offered as evidence that the adults of this species select as a spawning platform the leaf that is relatively easy to move through the water, since, while guarding their clutch of eggs, they consistently move the egg-carrying leaf when disturbed (Keenleyside and Prince, 1976).

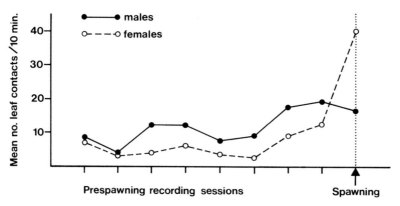

**Fig. 4.7.** Frequency of leaf contacts (nipping, tugging, and pushing) per 10 min by male and female *Aequidens paraguayensis* during the 48 h before spawning. (Modified after Timms and Keenleyside, 1975)

## 4.2.3 Eggs Sheltered

Many fish excavate a spawning site underneath or inside a structure on the substrate, thus modifying a natural shelter to increase the protection afforded their eggs. Objects used are rocks, ledges, empty mollusc shells and even man-made refuse. For example, the pomacentrid *Eupomacentrus leucostictus* spawns in excavations made inside empty conch shells, tin cans, drain pipes and holes in concrete blocks (Brockmann, 1973). A variety of motor patterns is used to create such spawning sites. The most common are:

*(a) Mouth-Digging.*The fish enters the enclosure, picks up sand, small pebbles, pieces of coral, shells, etc., in the mouth, carries them outside the shelter and spits them out. Provided the fish can lift the material in its mouth, this appears to be a simple, efficient method of excavating.

*(b) Pushing.*If an object is too large to be picked up, the fish may place its open mouth against it and swim vigorously forward, pushing the object away. The effectiveness of this technique depends on how deeply the object is imbedded in the substrate, and on its size relative to the fish.

*(c) Pectoral Fin-Digging.*Loose material can be moved out of an excavation by the pectoral fins. The male sculpin, *Cottus gobio*, combines vigorous alternate beats of its large pectoral fins with slight body undulations to move fine materials out of its spawning enclosure (Morris, 1954). The marine goby, *Bathygobius soporator*, uses a pattern called "waving" in which the body and pectoral fins oscillate at about six to eight beats per second, sending clouds of sand away from the digging fish (Tavolga, 1954).

*(d) Body- and Tail-Digging.*Another common technique is for a fish to settle onto the bottom and with rapid, vigorous tail beats and body undulations to plough forward, scattering the substrate materials to each side as it goes. The tail provides

76

the propulsive force and the head, anterior part of the body, and pectoral fins create a furrow in the bottom. Repeated digging at the same location produces a deepening excavation with loose materials being forced out of the enclosure. Some fish alternate between this and mouth-digging, as for example the pomacentrid *Abudefduf zonatus* (Keenleyside, 1972a).

The ability of fishes to use a variety of excavating techniques to suit the particular situation is well documented by the marine gobiid *Gobius microps* (Nyman, 1953). The male prepares a spawning excavation on sandy bottom under a single empty shell, usually of the clam *Mya arenaria*. If the concave side of the shell is upward, the fish turns it over. He then moves to the tapering end of the shell and tail-digs vigorously. This creates a shallow depression into which that end of the shell settles, and is gradually covered with sand. Meanwhile the other, more rounded end is raised off the bottom, creating a slight enclosure underneath. The male enters this and vigorously digs, sending sand out from under the shell and creating an excavation that eventually is large enough for male and female to enter and spawn.

Species that excavate enclosed spawning sites thereby create shelters in which the developing progeny can be efficiently protected against predation. Parental attention is also required to prevent the excavation from filling in with silt or sand, especially in inshore, marine areas, where tidal currents regularly shift large amounts of substrate materials. Silting-in is prevented by the guarding parent regularly fanning with its fins inside the enclosure, and by occasional digging.


## 4.2.4 Eggs Carried Away

Many mouth-brooding cichlid fishes create excavations that are used solely for courtship and spawning. When the brooding parent has finished spawning it leaves the area, carrying the eggs in its mouth. The various digging patterns described in the previous section are also used by these cichlids, although the patterns of some species have been examined in even greater detail.

For example, Barlow and Green (1970) recognized three distinct mouth-digging patterns in *Sarotherodon melanotheron* held in aquaria with gravel substrate. These were called sucking, scooping, and plunging. In *sucking* the fish pitches slightly head down and swims forward close to the bottom with mouth open, sucking up gravel which it spits out at the lip of the excavation. *Scooping* is similar but more like bulldozing; the lower jaw is pushed into the substrate, and the head and body undulate as the fish moves forward, scooping up gravel which it then spits out. *Plunging* is more direct. The fish pitches into a head-down vertical posture and with open mouth accelerates into the bottom. It then backs out with a mouthful of gravel and spits it out.

Cichlids, like the centrarchids described earlier, are also capable of clearing stones, rooted plants, and a variety of debris from a spawning-site excavation by using the mouth. For example, male *Hemihaplochromis philander* remove plants, roots and stones from a nest site by a variety of pushing, pulling and jerking movements (Ribbink, 1971) and when male *Sarotherodon andersonii* dig nests

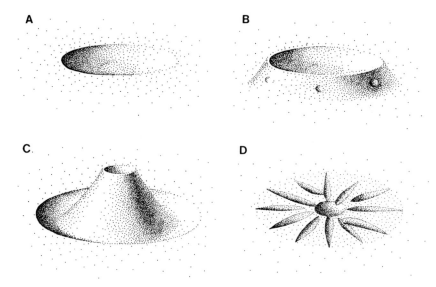

**Fig. 4.8.** Nests built by *Sarotherodon* males. **A** *S. andersonii*, **B** *S. variabilis*, **C** *S. macrochir* (Zambesi form), **D** *S. macrochir* (Congolese form). (Modified after Mortimer, 1960; Fryer and Iles, 1972)

among thick vegetation they bite off or pull out plants with the mouth (Mortimer, 1960).

An interesting feature of cichlid reproduction is the great range in size and complexity of nests created by different species in the large mouth-brooding genus *Sarotherodon* (Fig. 4.8). These are female mouth-brooders and the nests are made by the males as sites for both courtship and spawning. *S. andersonii* excavates a simple, shallow, saucer-like depression (Fig. 4.8A). *S. variabilis* males occasionally excavate by mouth several small pits around the periphery of the main nest (Fig. 4.8B). The function of these accessory pits is not known, but a possible explanation is provided by observations of *S. grahami* made by Albrecht (1968). This fish makes a small depression in the substrate and then nibbles at the inner sides of the excavation, creating several pits that are gradually enlarged until they combine to form a continuous ring or terrace around the central pit. Thus, small accessory pits seen around a central excavation may in some cases be an intermediate phase in the enlargement of a nest.

*Sarotherodon macrochir* has several distinct populations, whose taxonomic position is not clear (Mortimer, 1960; Fryer and Iles, 1972). Figures 4.8C and D illustrate nests created by the males of two populations. The Zambezi fish build a conical, volcano-shaped structure, with a small concave spawning platform at the top (Ruwet, 1962). The Congolese fish build one of the most elaborate of all cichlid nests (De Bont, 1949; Ruwet, 1963). It varies in diameter from 50 to 300 cm, and has a low, concave-topped spawning platform at the centre. From 6 to 12 grooves extend from the central mound to the nest perimeter, giving the entire structure the appearance of a spoked wheel. The grooves are made when the male settles to the bottom at the centre, pushes his open mouth into the substrate and swims vigorously to the periphery, ploughing a groove through the sand as he goes.

78

Several small pits may also be excavated around the nest centre, at the inner ends of the grooves (Ruwet, 1963). The functional significance of the grooves and pits is not clear, although the appearance of the nest gives the strong impression that the central spawning platform is made more conspicuous by the radiating grooves. Thus the form of the nest may combine with male courtship behaviour to attract ripe females to the spawning site.

## 4.3 Nest Built of Collected Materials

Some fishes prepare for spawning by gathering materials together to form a nest, into which the eggs are placed. The nest is guarded and repaired if necessary by one or both parents until the eggs hatch and the young disperse. Here I consider two groups of fish that follow this strategy: the bubble-nest builders (Belontiidae) and the sticklebacks (Gasterosteidae).

### 4.3.1 Belontiidae

These are "air-breathers", or labyrinth fishes, so-called because they have a suprabranchial, labyrinthiform, accessory respiratory organ that allows them to acquire oxygen from air taken in at the water surface. With this respiratory adaptation they can inhabit waters that are shallow, weedy, and warm, and where the oxygen content may be very low (Forselius, 1957).

Associated with this habitat and method of acquiring oxygen is a highly specialized form of nest-building in some species. Typically a mature male makes a floating nest of air bubbles that are produced in his mouth. He sucks in air at the water surface and creates small bubbles surrounded by mucous secreted by special buccal glands. The bubbles are spit out at one site to form the nest. Sometimes a nest is built around emergent vegetation, serving to anchor it. The male usually builds and maintains the nest, but in some species the female occasionally contributes (Braddock and Braddock, 1959; Miller and Robison, 1974). Spawning occurs in mid-water, below the nest. The male picks up the fertilized eggs and spits them into the nest, where they remain floating at or near the surface.

Pieces of vegetation are often incorporated into the nests of belontiids, presumably to add strength and durability. These may be fragments collected at the water surface, or pieces bitten or pulled from rooted plants. Some species occasionally add detritus, sand, and even faeces to their nests. In *Colisa lalia* the initial stage of a nest is often a thin layer of collected plant fragments to which the male quickly adds air bubbles. In general, nests can have a variety of shapes and sizes in both vertical and horizontal dimensions, and the amount of plant material varies according to its availability and the species of fish (Forselius, 1957; Miller and Robison, 1974).

Consolidation and maintenance of the nest is achieved in several ways. The male may shift bubbles and plant fragments from one location to another, as though trying to maintain a more or less uniform, spherical structure. Occasionally he orients vertically under the nest and butts repeatedly with his mouth against the under surface. This may release buccal gland mucous that contributes to better adhesion among the nest components and eggs (Forselius, 1957). Throughout the building and maintenance of his nest the male frequently orients towards it from beneath, indicating that he visually inspects the structure, and then builds or shifts materials according to its condition. Bubble-nest building is clearly an active, dynamic process that continues throughout the development of eggs and newly hatched young.

### 4.3.2 Gasterosteidae

The reproductive behaviour of sticklebacks has been studied more intensively than that of most other fishes, thanks largely to the detailed laboratory investigations of the three spine stickleback (*Gasterosteus aculeatus*) extending over several decades by Tinbergen and his associates (Tinbergen and van Iersel, 1947; Tinbergen, 1951; van Iersel, 1953; Sevenster, 1961; van den Assem, 1967). A recent monograph by Wootton (1976) summarizes published information on the biology of the family.

It is a small group, containing three monospecific genera (*Apeltes, Culaea*, and *Spinachia*) and two genera (*Gasterosteus* and *Pungitius*) with two species each. The taxonomic status of the latter two genera is currently in doubt; they have very wide distributions and more species may eventually be recognized (Wootton, 1976). There is one strictly marine species (*S. spinachia*), and the others are anadromous or entirely freshwater.

The breeding system is similar in all species. The sexually mature male establishes a territory on or near the substrate, within which he builds a nest. He then attracts a ripe female by distinctive courtship behaviour. Eggs are laid and fertilized inside the nest, and the male guards them until they develop into fry and disperse. Promiscuous mating is universal in the family; pair formation is restricted to the brief period of courtship and spawning.

The factors influencing nest-site selection in *G. aculeatus* have been studied experimentally by a number of workers (van Iersel, 1958; van den Assem, 1967; Jenni, 1972). The results were reviewed by Wootton (1976). Briefly, the presence of rooted vegetation and of nest-guarding, conspecific males are both important. In small aquaria, with sandy bottom and either a row of plants, a single plant, or both, solitary males tended to build their nests away from the plants, but not at the maximum distance possible. In tanks up to 6 m long, a tendency to nest near or even within a row of plants was more pronounced, especially when the plants were well away from the tank walls. It appears that fish in a tank containing just a few plants select a nest-building site near the plants, and that this tendency is stronger when the open areas are larger. Presumably the plants offer support and shelter for the nest, and hiding places for the male. The few available records on nest locations in free-living *G. aculeatus* support this conclusion; nests are usually built on open substrate near shelter (Wootton, 1976).

The presence of other males also affects nest site selection. When several males were placed in a large tank at the same time, their nests were built on average closer together than when they were introduced serially. Breeding male *G. aculeatus* are highly aggressive, and males that establish nests simultaneously seem to have more equivalent aggression levels than do those nesting farther apart in time. In the latter case a newcomer is at first much less aggressive than any residents, and hence builds his nest relatively far from them. Eventually, if more and more males enter an area to breed, the nests become more uniformly spaced, resembling the pattern that is often seen in nature (van den Assem, 1967).

The ninespine stickleback (*Pungitius pungitius*) has two distinct forms in North America, the Bering and the Mississippi (McPhail, 1963). In a comparative study of these two and the common European form of the same species Foster (1977) placed males in a large "naturalistic environment" tank, with substrate having many small rock piles and clumps of *Elodea*. Ninety-five percent of Mississippi males built nests on bare rubble substrate, while 100% of the other two forms built nests in the plants. These choices reflect the known nesting habitats of the different forms in nature (Morris, 1958; McKenzie and Keenleyside, 1970).

*Nest-Building Behaviour.* Stickleback nests are built either on the substrate, usually in a shallow pit, or close to it, attached to rooted aquatic plants. Substrate-nesters are *G. aculeatus* and *G. wheatlandi*; plant-nesters are *Culaea inconstans, Apeltes quadracus* and *Spinachia spinachia*. *P. pungitius* nest-sites are variable; in some habitats nests are built in substrate pits, in others they are attached to plants. Nest shape in the family varies from spherical to elongate, and the nest includes a central opening which may be a shallow, cup-like depression, a deeper pit or a complete tunnel open at both ends. Eggs are placed inside the central cavity. The materials used in nest construction depend on the characteristics of the breeding habitat. Most studies of stickleback reproductive behaviour have been done with aquarium-held fish, provided with nesting materials selected by the investigator. Direct observations of breeding sticklebacks in the field are uncommon, but suggest that species with wide distribution (e.g., *G. aculeatus* and *P. pungitius*) use whatever nesting materials are locally available.

The repertoire of motor patterns used to construct nests is extensive. The following comparison is based mainly on observations of aquarium-held fish, and includes the six best-known species. The available information is summarized in Table 4.3.

*Mouth-Digging.* Species that built nests on the substrate begin by excavating a shallow pit. Fine materials are moved by plunge-digging. Plant rootlets, small stones, etc. are carried away in the mouth.

*Collecting Nest Materials.* All species use the same behaviours to collect materials for nest construction. These include: *searching*, in which the male swims actively about, frequently stopping and visually fixating potential nest material; *breaking-off* pieces of plant by biting, jerking, pulling, or twisting; *testing*, in which he picks up a plant fragment in the mouth, spits it out, takes it in again, finally either dropping it or; *carrying* material to the nest site in the mouth. Typical materials used are fragments of plant leaves, rootlets, and strands of filamentous algae.

Morris (1958) examined the preferences of male *P. pungitius* and *G. aculeatus* for nest materials of different colours by offering them pieces of cotton of different

**Table 4.3.** Presence (+) or absence (−) of nest-building motor patterns in male gasterosteids of six species

| Species | Mouthdigging | Collecting materials | Glueing | | Pushing | Boring | Excavating below nest | Carrying sand to nest | Tunnelling | Nest extension |
|---|---|---|---|---|---|---|---|---|---|---|
| | | | Superficial | Insertion | | | | | | |
| *Gasterosteus aculeatus* | + | + | + | − | + | + | + | + | + | + |
| *G. wheatlandi* | + | + | + | − | + | + | + | + | + | + |
| *Pungitius pungitius* | ± | + | + | + | + | + | ± | − | + | + |
| *Culaea inconstans* | − | + | + | + | + | + | − | − | + | + |
| *Apeltes quadracus* | − | + | + | − | + | − | − | − | − | + |
| *Spinachia spinachia* | − | + | + | − | + | + | − | − | ? | ? |

hues or of different shades of grey. *P. pungitius* tended to select the lightest-coloured threads and Morris suggested that in nature this tendency would lead to selection of relatively young pieces of plant that would live longer and decompose more slowly than older, more mature pieces. The only preference shown by *G. aculeatus* males was for red threads in the late stages of nest construction, resulting in a conspicuous entrance to the nest tunnel. Nests of this species are often built on open substrate with few local landmarks, and a conspicuous nest entrance might therefore serve as a guide for the male and for females when approaching the nest (Morris, 1958). If this argument is valid, it must be remembered that enhanced visibility of the nest will also make it more vulnerable to predation.

When suitable nest materials are scarce, nest-building males may collect materials by "stealing" them from another stickleback's nest (Morris, 1958; Wootton, 1971). However, since male *G. aculeatus* will also remove eggs from another male's nest, either to eat them or to place them in their own nests (van den Assem, 1967), the possibility exists that some apparent "stealing" of nest material is actually "egg-stealing".

*Glueing.* Several distinct activities are used to form collected materials into a cohesive nest. Glueing involves the secretion of a substance from the male's kidneys, which, on contact with water, becomes relatively hard and sticky. All stickleback species use this "glue" to bind together the fragments of materials brought to the nest site. Two forms of glueing were recognized by Morris (1958): superficial and insertion glueing.

*(a) Superfical Glueing.* The male secretes a long, thin thread of glue onto the outer surface of the nest. In species that build nests among plants, the male may wind the thread one or more times around the nest and the adjacent plants, thus serving to anchor the nest to its supporting structures. *C. inconstans* may secrete glue at the presumptive nest site before any nest materials have been collected, thus providing a sticky substrate onto which the initial materials can be placed (McKenzie, 1964). All species perform superficial glueing.

*(b) Insertion Glueing.* While hovering in front of his nest, the male bends his body in an arc, rotates slowly in the horizontal plane and releases a small spherical drop of glue or a thick strand up to 2 or 3 cm in length. He quickly picks up the glue in his mouth and deposits it firmly against the inner lining of the nest (Morris, 1958). Insertion glueing undoubtedly contributes to the solidity of the nest. Only *P. pungitius* and *C. inconstans* are known to perform this behaviour.

*Pushing.* This is common to all species. The male presses plant materials into the interior or onto the outside of the nest with his snout. Newly collected plant fragments, loose ends of previously incorporated fragments and pieces dislodged by nest building activities are all pushed back into the nest. The function is clearly to consolidate the nest.

*Boring.* This is a vigorous pushing of the snout into the collected mass of plant fragments. It results at first in a shallow depression that is gradually deepened and broadened into a tunnel, in which the eggs are later laid. The tunnel is usually horizontal in substrate nests, but may be in any plane in nests fastened to rooted plants above the substrate. Boring presumably serves the same basic function as pushing; loose pieces of nest material, together with glue, are consolidated into a well-knit, cohesive nest. In addition, it creates a tunnel for holding the eggs.

A. quadracus is the only species that does not perform boring. The interior of its nest is an open depression rather than a tube-like tunnel, and while the male pushes nest materials inward to form this depression, the pushing movements do not develop into vigorous, full-scale boring (Rowland, 1974a).

*Excavating Below the Nest.* In the substrate-nesting species the male may remove sand from below the nest by thrusting his snout deeply into the periphery of the nest, picking up sand and then releasing it to fall onto the nest surface. This may be repeated at several locations, resulting in a number of small pits around the edge of the nest, and a gradual accumulation of sand on top. The function seems to be to consolidate the structure by pushing the plant fragments deeper into the substrate and partially covering them with sand.

*Carrying Sand to the Nest.* This is shown only by the two *Gasterosteus* species. The male picks up sand in his mouth and drops it over the top, and especially around the entrance to the nest. Eventually the entire nest may be covered. This has two main functions: to weigh down, and thus consolidate and strengthen the entire structure; and to camouflage the nest which, in these species, is often in a somewhat exposed location (van Iersel, 1953).

*Tunnelling.* The male enters the central cavity of his nest, usually performs a bout of boring, then pushes on through the nest, breaking through the blind end and thus forming a complete tunnel. In *G. aculeatus* this act has been called "creeping through the nest" and appears to mark a transition between the nest-building and courtship phases of the male reproductive cycle (van Iersel, 1958; Nelson, 1965). Since tunnelling was rarely observed in *G. wheatlandi*, and was not in a clear temporal relationship to courtship behaviour, McInerney (1969) considered it to be simply an act of high intensity boring.

In *C. inconstans* the male does not tunnel through the nest, although the female does after laying a clutch of eggs in the nest (McKenzie, 1964; Reisman and Cade, 1967). Tunnelling does not occur in *A. quadracus*, since neither male nor female passes directly through the nest at any stage in the breeding cycle (Rowland, 1974a).

*Nest Extension.* In all stickleback species spawning is usually followed by a burst of nest-building activity by the male, consisting mainly of boring, pushing and glueing. This repairs any damage done to the nest by the vigorous actions of the two spawning fish. As a result the nest is strengthened, and often also extended in size; thus if several clutches of eggs are laid successively, the nest is gradually made larger. In *G. aculeatus* the male fastens a newly fertilized clutch firmly into the bottom of the nest by vigorous vertical boring movements, and then brings fresh nest material to the entrance and incorporates it there into a forward extension of the nest. These successive additions tend to overlap each other, much like roof tiles (van Iersel, 1953).

Elaborate nest extension behaviour is also shown by *A. quadracus*. After the first clutch of eggs is laid in his cup-shaped nest, the male extends it upwards around and over the eggs. The upper part of this extension is then shaped into a concave surface, ready to receive a second clutch of eggs. The same sequence may be followed several times, resulting in the multi-tiered or "apartment-like" nest often found in this species (Rowland, 1974a). Figure 4.9 shows three successive stages of such a multi-clutch nest. Nothing is known about nest extension in *S. spinachia*, although males are promiscuous, as in all other sticklebacks, and can maintain clutches of eggs from several females in a single nest (Sevenster, 1951), suggesting that repair and extension building may be necessary on occasion.

*Comments on Stickleback Nest-Building Behaviour*. As Table 4.3 indicates, the substrate-nesters have a richer repertoire of nest-building motor patterns than do those species that attach their nests to plants. Most of this difference is the result of the substrate-directed actions of the former group, i.e., digging an excavation, and periodic excavating below and carrying sand to the nest. *P. pungitius* that nest in plants show none of these three patterns.

However, an important feature of stickleback nesting not illustrated by such a summary table is the effect on nest-building behaviour within a single species of variation in breeding habitat, and hence in materials available for building. This has often been demonstrated in the simplified environment of an aquarium. For example, *A. quadracus* will build a nest of algae, detritus and sand directly on the bottom of a tank without rooted vegetation (Rowland, 1974b). They will also attach their nests to different plant species and even directly onto plastic air tubing (Fig. 4.9). *P. punigitius* from one Lake Huron population will spawn in barely perceptible algal nests under rocks in an otherwise bare tank, but make large, loose-fitting nests of algae attached to rooted plants when those materials are present (McKenzie and Keenleyside, 1970; Foster, 1977).

Field observations are still more revealing. *P. pungitius* build elaborate nests firmly attached to plants in quiet, weedy waters (Morris, 1958), simple algae-fragment nests under stones on rocky, wave-swept lake shores (McKenzie and Keenleyside, 1970), and nests at the bottom of burrows in soft, organic mud substrate (Griswold and Smith, 1972). *G. aculeatus* nests have been found on a wide variety of substrates, including mud, silt, sand, algae tufts, and crevices in rocks (Hagen, 1967; Black and Wootton, 1970; Wootton, 1971). Clearly, nest-building is a relatively plastic behaviour, at least in these two species. This plasticity must be partly responsible for their widespread distribution (Wootton, 1976).

In spite of the great popularity of *G. aculeatus* in ethological research, there is a dearth of published comparative work on stickleback behaviour. Detailed comparisons of nesting and other reproductive behaviours, both among species and among geographically distinct populations of one species, would lead to a better understanding of the role of behaviour in the overall adaptation of these species to their environments.

*Summary Comments*. The active selection and preparation of a spawning site appears to be limited to externally fertilizing species with demersal eggs that provide some post-spawning care for their progeny, either directly, through a

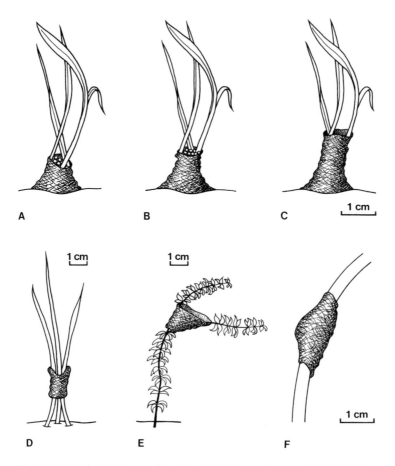

**Fig. 4.9.** Variation in form and location of *Apeltes quadracus* nests. **A,B,C** three stages in construction of one nest as successive egg clutches are laid, **D-F** nests at different locations on plants, and on plastic air tubing. (After Rowland, 1974a)

variety of parental care behaviours, or indirectly, by covering their eggs before abandoning them. Species with buoyant, pelagic eggs do not require a platform for egg deposition, and therefore show no spawning site preparation.

However, not all demersal spawners actively prepare the substrate to receive newly fertilized eggs. For example, in the large freshwater family Cyprinidae, several species use a scatter-spawning technique. Groups of adults move into shallow, weedy water and release their gametes simultaneously, with the eggs settling among the vegetation. No further protection is given the eggs by their parents. Yet it seems highly likely that the adults exercise some choice in selecting a spawning site. Eggs broadcast onto open, sandy substrate, for example, would be more vulnerable to predators than those scattered among thick stands of aquatic plants. Thus, the adaptive significance of the various pre-spawning activities of fishes cannot be evaluated accurately without detailed knowledge of the entire reproductive process, including courtship, spawning and parental care.

86

Chapter 5

# Breeding Behaviour

Breeding in fishes consists of three successive stages: courtship, mating and spawning. These are defined as follows: *Courtship* is the heterosexual reproductive communication system leading up to mating (Morris, 1956; Baylis, 1976b). *Mating* is the sexual act itself in which at least one female and one male come close together and release their gametes, either more or less simultaneously into the surrounding medium (external fertilization), or by the transfer of sperm from the male into the female (internal fertilization). *Spawning* is the release of gametes or of developing young to the external environment. It is virtually synonymous with mating in species with external fertilization, but where fertilization is internal, the release of young occurs at some post-fertilization stage.

In this chapter I begin with a brief discussion of the functions of courtship behaviour and then describe the various breeding systems, organized according to the methods used by fishes to protect their newly fertilized eggs. It is by no means an exhaustive survey; several of these are already available (e.g., Breder and Rosen, 1966; Sterba, 1966).

I have selected certain species to illustrate both the diversity and the similarities in fish breeding systems, with the aim of identifying general patterns of breeding adaptation to diverse environments. Therefore, wherever possible, species used to illustrate each system are taxonomically diverse and breed in different habitats. The most obvious limitation on this sampling is that information on breeding behaviour is available for only a small proportion of known fishes. These tend to be either of economic importance or easy to breed in captivity.

## 5.1 Functions of Courtship Behaviour

Animal courtship appears to serve several distinct functions (Tinbergen, 1953; Morris, 1956; Forselius, 1957; Bastock, 1967; Barlow, 1970; Baylis, 1976b). These can be grouped conveniently into three main categories:

(a) Attraction and identification — so that sexually mature males and females of the same species are brought together at a place and time appropriate for breeding.

(b) Arousal, appeasement and synchrony — in which the motivational states of males and females are fine-tuned so that the appropriate breeding motor patterns are synchronized and successful mating occurs.

(c) Longer-term effects, such as the establishment and maintenance of a firm pair bond in species where the breeding male and female collaborate in caring for their young.

### 5.1.1 Attraction and Identification

Various mechanisms operate among fishes to ensure the presence of mature male and female conspecifics together on the spawning grounds. For more or less permanently schooling species (e.g., some clupeids) attraction is no problem; both sexes travel in the same school. Many other species (e.g., among the cichlids, centrarchids, pomacentrids) establish breeding territories, usually occupied by males. Courtship behaviour by these residents attracts potential mates.

Mate attraction depends on an exchange of signals, i.e., on a communication system. Such systems have often been described as a series of alternating stimulus–response reactions between male and female. The best-known ethological example is probably the reaction chain of courting threespine sticklebacks, *Gasterosteus aculeatus* (Tinbergen, 1951). Courtship by a pair of animals seldom proceeds directly through such an alternating chain of events; often it is disrupted by one individual failing to respond to a particular signal in an appropriate way. Sometimes one or more links are left out as a highly aroused animal moves quickly to the final stages. Generally, as the motivational states of the two animals become synchronized, courtship proceeds further and further, until mating is attempted (Morris, 1956; Forselius, 1957).

The exchange of information between potential mates can be based on different sensory modalities. In fishes, most work on courtship communication has centred on visual signals, especially those involving movements, postures, and changes in colour patterns. This is partly a function of the relative ease with which observers can document visual signals, as opposed to those of other modalities. However, fishes have highly developed chemosensory (Hara, 1971; Kleerekoper, 1969) and acoustic (Tavolga, 1971) systems, and these may also be involved in courtship signalling.

One presumed function of courtship behaviour is the prevention of hybridization between closely related species. Ethological isolating mechanisms are among the most important barriers to interspecific breeding (Mayr, 1963), and the successive exchange of signals between a courting male and female provides the opportunity for correct identification of conspecifics. The signals are more generalized at early stages in courtship, becoming more discrete and species-specific at later stages. It is this increasing specificity that militates against continued courtship by individuals of different species.

### 5.1.2 Arousal, Appeasement, and Synchrony

The exchange of information between a courting male and female promotes their mutual sexual arousal, culminating in the release of gametes. For externally fertilizing species (the vast majority of fishes) this release must be simultaneous or nearly so, because the gametes are capable of fertilization for only a short period after release. Among internally fertilizing species the male must transfer sperms into the female's genital tract; this also requires a high degree of synchrony since the transfer usually occurs during a very brief moment of contact (Clark et al., 1954).

However, courting fishes that also have breeding territories are often highly aggressive. Appeasement, or reduction of this aggression, so that the two individuals can remain close together during spawning, is presumed to be another function of courtship (Tinbergen, 1953). The interrelationships between arousal and appeasement during courtship in cichlid fishes has been examined in detail by Barlow (1970) and Barlow and Green (1970).

Bastock (1967) discussed the relative levels of courtship activity by males and females and concluded that, in general, either both fish are equally active or the male is the more active partner. However, even among internally fertilizing species such as the guppy, *Poecilia reticulata*, where the colourful, smaller male is clearly much more active and conspicuous during courtship than the larger, more drab female, the latter is far from passive. Active cooperation and the performance of specific postures by the female are required for successful copulation (Liley, 1966), and this probably applies to most species. In the mouthbrooding cichlid *Sarotherodon melanotheron* the smaller of two courting fish, regardless of sex, performs more active courtship. Since larger size is also correlated with dominance, the greater courtship activity by the smaller member of the pair supports the appeasement function of courtship behaviour (Barlow and Green, 1970).

## 5.1.3 Long-Term Effects

Fishes in which a male and female establish a pair bond and collaborate in all phases of their reproductive cycle, including extended care of the young, must maintain a stable within-pair relationship to be successful in breeding. One function of courtship may be to promote the establishment of such a firm pair bond. This could explain the lengthy period of courtship before spawning in many biparental cichlid species (Baylis, 1976a), where repeated reproductive cycling by the same pair is common, at least in aquaria. Little is known about the duration of pair bonds in nature, although in 20 m × 10 m outdoor ponds we have observed the same pairs of *Herotilapia multispinosa* breeding repeatedly over three months (Keenleyside, unpublished data).

It would be useful to have data from biparental species comparing the duration and intensity of pre-spawning courtship behaviour with post-spawning pair bond stability and reproductive success. If the relationship is a direct one, then clearly the courtship period has an important long-term function.

## 5.2 Mating with External Fertilization

### 5.2.1 Species with Pelagic Eggs

Most information on breeding behaviour of fishes with pelagic eggs comes from studies on coral reefs. However, one of the world's most important food fishes, the Atlantic cod (*Gadus morhua*), produces pelagic eggs, and its reproduction has been studied in detail in aquaria (Brawn, 1961).

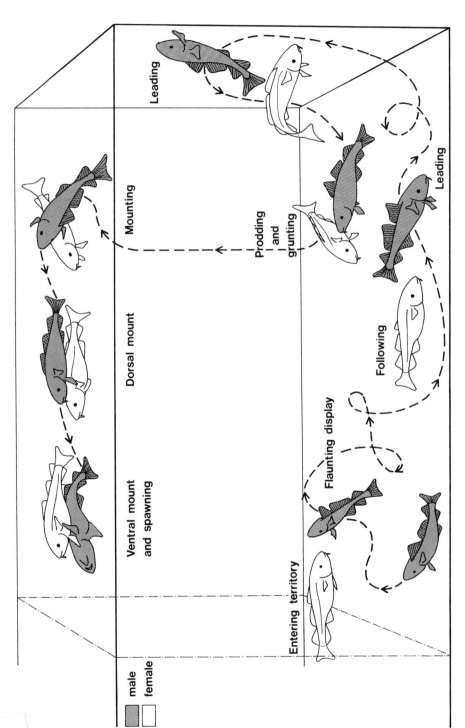

**Fig. 5.1.** Courtship behaviour in *Gadus morhua*. See text for details. (After Brawn, 1961)

*Gadidae*. Cod spawn in large aggregations from near the surface to a maximum depth of 200 m (Leim and Scott, 1966). Females produce very large numbers of buoyant eggs; a 1 m long fish releases about 5 million eggs (Powles, 1958). Despite the proximity of many mature adults, synchronized, indiscriminate spawning is not practised. Rather, breeding occurs in pairs following elaborate courtship (Brawn, 1961). Figure 5.1 summarizes the events, as seen in a large tank (11 m³).

A mature male establishes a territory and begins courtship when a ripe female approaches. He performs a "flaunting display" in which all median fins are fully spread and the body makes wide amplitude lateral bends directly in front of the female. He then swims in loops along an undulating path away from her and eventually towards the water surface. If she does not follow he returns, repeats the flaunting and looping movements, produces loud grunting sounds, and may prod the female with his snout. Eventually both fish reach the surface, where the male moves onto the female's back, presses his lower jaw onto her head and his pelvic fins against her sides. This is the "dorsal mount". He then slides around sideways, still holding her with his pelvic fins, until he is in an inverted position below her ("ventral mount"), and the genital openings of both fish are close together. Eggs and sperm are released as both fish beat their tails and move in a horizontal circular path. Thus, cod spawning follows a complex series of interactions between male and female, involving visual, tactile and vocal signals.

*Scaridae (Parrotfishes)*. An understanding of breeding behaviour in this family is complicated by the almost universal existence of protogynous hermaphroditism (Robertson and Hoffman, 1977). Some species are *monandric* (one male type only); all individuals begin life as females, and some change into males as they age. Others are *diandric*: some fish begin life as males, and remain so throughout life (these are called primary males), and others begin as females and later change sex (secondary males). In general, four sexual identities have been recognized among the scarids: females, primary males, secondary males, and transitional fish that are changing from female to male (Reinboth, 1970).

Associated with this variation in sexual identity is a marked colour dimorphism among scarids. In many species the largest fish are distinctively and brigtly coloured ("gaudy"), and are all males, either primary or secondary. Other adults are smaller more plain-coloured ("drab"), and are mostly female, but in most species include some primary males (Choat and Robertson, 1975).

Two general breeding patterns occur among scarids: pair-spawning, between one male and one female; and group-spawning by three or more individuals. In species with gaudy large males, it is these fish that pair-spawn with females. Group-spawners are usually all drab, although in *Scarus globiceps* gaudy males occasionally join a spawning group (Choat and Robertson, 1975). Group-spawning is as follows. A few fish (from 3 to 15) suddenly leave a larger aggregation, dash diagonally upward a short distance, release gametes and dive back to the larger group. Underwater filming of group-spawning in *Sparisoma rubripinne* showed a close synchrony of gamete release by all fish at the peak of the upward run (Randall and Randall, 1963).

Spawning by a single pair of scarids is preceded by elaborate, conspicuous courtship. In *Scarus croicensis* some large, gaudy males establish territories near the substrate. As a group of drab conspecifics approaches, one or more of the gaudy males performs a "bob-swim display" near the group. The displaying male has dorsal, anal, and caudal fins folded and moves along a sinuous, rising and falling path. If he is joined by a female from the group the pair dashes rapidly and diagonally upward, turns sharply and descends, the fish separating as they dive. Eggs and sperm are presumably shed at the apex of the climb (Barlow, 1975). *Sparisoma aurofrenatum* has been observed spawning in pairs, in which the two fish rotated around each other as they swam rapidly towards the surface, then released gametes and immediately descended (Winn and Bardach, 1960).

In their extensive study of over 20 scarid species at Heron Island, Great Barrier Reef, Choat and Robertson (1975) only observed spawning (both group and single pair) at the reef crest or on the outer, seaward slope of the reef, and at or just after high tide, when strong currents were running in the area. Thus spawning by these fishes appears to be timed so that newly fertilized eggs are carried away quickly in the plankton.

*Labridae (Wrasses)*. Tropical labrids produce pelagic eggs, and both pair- and group-spawning are known to occur in some species. In temperature waters some labrids lay demersal eggs in nests built and guarded by adult males (Fiedler, 1954; Potts, 1974); these species are not considered here.

The mating system is variable among tropical labrids. The cleaner wrasse, *Labroides dimidiatus*, lives in stable groups, or harems, consisting of a single adult male and five or six females of varying size. The male spawns regularly with each of his females. Spawning is preceded by courtship displaying by both sexes and leads to the coordinated upward spawning rush typical of other pair-spawning labrids and scarids (Robertson and Hoffman, 1977). The most striking feature of the *L. dimidiatus* harem system is that when the male disappears the largest female (who is dominant over the other females) immediately begins transforming into a male, takes over defence of the entire harem area, and soon shows male sexual behaviour towards the other females (Robertson, 1972).

*Thalassoma bifasciatum*, the common bluehead wrasse of western Atlantic coral reefs, presents a more complicated picture (Randall and Randall, 1963; Reinboth, 1973; Warner et al., 1975; Robertson and Hoffman, 1977). Females and small males have several different colour patterns; the large males are much gaudier, with a blue head and green body separated by a white collar bordered with black. Each day about noon the largest males establish territories in an arena on the outer, downcurrent edge of the reef. Smaller, non-territorial males gather just inshore from the territories. Females approach and try to pair-spawn with the largest males. A vigorous courtship precedes these spawnings. A territorial male swims vertically upward about 1 m, quivering his tail and body, then turns and dives to his starting position. These vertical excursions become more frequent as a ripe female approaches. He then circles above her along an upward spiral path with his tail quivering. If the female follows, the male slows and the two fish suddenly dash upward with the female slightly in front. She turns so that genital openings of both fish are close together, he bends his body sharply, they spawn, separate, and descend. This description is based on observations and filming by Reinboth (1973).

92

The smaller males usually mate in a group of several males and one female, following the pattern described above for group-spawning scarids (Randall and Randall, 1963; Reinboth, 1973). However, they also occasionally try to spawn as individuals with mature females, either by dashing into a large male's territory and joining him as he rises to the surface with a female, or by tactile stimulation of a female as she approaches a territorial male, thus apparently trying to initiate a spawning rush with her. Since large territorial males may spawn 40 or more times per day, while the smaller males spawn much less often (whether in groups or pairs) it is clearly advantageous for males to become large, brightly coloured and territoral (Warner et al., 1975).

In *Thalassoma lunare*, studied at Heron Island, group-spawning occurs among large aggregations of drab fish, most of which are males. A small group, apparently composed of one ripe female and up to several dozen males, begins to rise above the aggregation. All fish jostle and push against one another and the female appears to be in the lead. Then the typical fast, steep, spawning run occurs, at the peak of which a large cloud of milt appears, presumably from several males, and the group descends. Pair-spawning in this species occurs when a brightly coloured territorial male displays at an approaching female with body vibrations, short chases, and bursts of rapid pectoral fin fluttering. If the female continues to approach, the male continues vibrating, the two fish dart rapidly upward, turn abruptly, release eggs and sperm and then descend (Robertson and Choat, 1974).

*Comments.* The species producing buoyant, pelagic eggs use a spawning strategy that minimizes predator pressure on the unprotected eggs and the breeding adults. Although they are mainly substrate-oriented fishes, spawning occurs in open water near the surface, usually in fast-moving currents and when light intensity is falling. This reduces the vulnerability of eggs and adults to predation, while allowing the visual contact needed for the exchange of courtship signals (Barlow, 1974a; Robertson and Hoffman, 1977). Spawning is well synchronized, often occurring at the instant when the upward-moving fish come closest together. Probably both visual and tactile signals are used at the critical moment. There is no parental care; newly fertilized eggs simply join the rapidly drifting plankton. The relationship between dispersal of pelagic eggs and subsequent settling of the larval fishes, especially in regions where suitable coral reef habitat is extremely patchy, is discussed by Jones (1968) and Sale (1977).

The presence of both group- and pair-spawning in some of the coral reef labrids and scarids allows for variation in the mating system according to size and sex ratio of the local population. Where large, conspicuous, territorial males are relatively scarce, as in the harem-forming *L. dimidiatus* and arena-forming *T. bifasciatum*, pair-spawning may be the most efficient way to ensure fertilization of each female's eggs. Also, pair-spawning might be expected if the local population density is low, since a conspicuous, territorial male may be able to spawn even if few ripe females are present (Reinboth, 1973). Where large numbers of a species are present, and most are plain-coloured (both male and female), then group-spawning may be the most efficient breeding system, especially if the sex ratio of mature adults in the large group is heavily skewed towards males, as in *T. lunare*.

## 5.2.2 Species with Demersal Eggs, no Prolonged Guarding

Many fishes lay eggs that are heavier than water and sink to the bottom when released by the female. Protection from predation is provided by various strategies, an effective one being active guarding by one or both parents. However, parental care is well developed only in certain groups; many species with demersal eggs either simply release them onto the bottom, actively bury them in the substrate, or place them inside shelters. In all these cases, guarding by the parents is non-existent or brief.

### 5.2.2.1 Eggs Released onto Substrate

*Clupeidae (Herrings).* The herrings are pelagic, schooling fishes, some of which are anadromous, others strictly marine. Despite the great commercial importance of several species, little is known of the spawning behaviour of most (Leim and Scott, 1966; Hart, 1973). The Pacific herring (*Clupea harengus pallasi*) is an exception; it has been observed spawning in nature and in the laboratory. Herring spawn in large schools in or just below the intertidal zone of coastal British Columbia, where the eggs adhere to plants, rocks, wooden pilings, etc. (Hart and Tester, 1934). While spawning, the female turns on her side, extends her fins and, with her tail beating at increased frequency and decreased amplitude compared to normal swimming, releases strings of adhesive eggs as her genital papilla brushes over the substrate. The males release sperm, either while following a few centimetres behind a female (Rounsefell, 1930), while passing over the eggs a short time after the female (Hourston et al., 1977), or while swimming in the vicinity of spawning females (Schaefer, 1937). Apparently no courtship occurs.

Spawning individuals are usually in large schools and once the process begins it quickly spreads throughout the entire school. Introduction of herring sperm into a large tank containing ripe adults induced spawning by both sexes (Hourston et al., 1977), suggesting that chemical stimuli promote synchronous spawning. The milky sperm released by millions of fish at one location in nature (Hourston and Rosenthal, 1976; Outram and Humphreys, 1974) may discolour the water to such an extent that visual contact between fish must be reduced, again suggesting that chemical stimuli are important for spawning synchrony.

*Esocidae.* Northern pike (*Esox lucius*) breed in shallow water, over soft, mucky substrate along the marshy edges of rivers and lakes. The spawning unit is either a single pair or a group of two or three males with one female. The fish in pairs are about equal size; in groups, the males are usually smaller than the female (Clark, 1950), suggesting that the larger, older males are able to keep smaller males away from their spawning partner.

Spawning begins with the male gently rubbing his snout against the female's head and body. The two fish then swim forward together, the male begins lateral undulating movements and finally he strikes the female vigorously with a strong lateral tailbeat. At this point the genital pores of both fish are close together and the

94

gametes are shed. The eggs sink immediately to the bottom where they stick to debris and vegetation. At the moment of the male's lateral "mating thrust", his anal fin is slightly folded into a spoon shape. This probably directs his sperm towards the female's genital pore, thus facilitating immediate fertilization of the eggs (Fabricius and Gustafson, 1958).

*Cyprinidae.* Within this large freshwater family are many scatter-spawning species. A common breeding unit is several males with one female. The carp (*Cyprinus carpio*) is one of the best known of these species (Swee and McCrimmon, 1966; McCrimmon, 1968). Small compact groups move through marshes and weedy margins of lakes, in water less than 0.5 m deep, with the males repeatedly pushing their heads against the female's body. Finally she responds by raising and vigorously beating her tail, releasing eggs as she dashes forward. The males stay close beside her and lash their tails while releasing sperm. The adhesive eggs scatter onto the surrounding vegetation. These spawning groups are conspicuous in shallow, marginal waters, as they repeatedly splash at the surface while spawning.

A similar mating pattern, with one female pursued by several males, has been described for *Hybognathus nuchalis* (Raney, 1939), *Mylocheilus caurinus* (Schultz, 1935), and *Richardsonius balteatus* (Weisel and Newman, 1951). Typically, many adults of both sexes crowd together along lake margins or over stream riffles. Several males push against each other, while trying to stay close to one female. Then, with simultaneous vigorous body quivering, all fish in the small group spawn. Male *M. caurinus* try to place their tails over the female's caudal peduncle at the moment of spawning, presumably to enhance rapid fertilization (Schultz, 1935).

The single male–female pair is also common among breeding cyprinids. Within the large Asian genus *Barbus*, several species are known to spawn in pairs in aquaria (Kortmulder, 1972). Breeding begins when a male pauses beside a female with his body in a shallow C-curve, and leans towards her. He may then swim away in front of her towards a clump of vegetation with his tail beating at smaller amplitude than in usual swimming. He may also circle around her while C-curving and leaning. Spawning occurs when the male moves close beside the female, curves his tail over her back, and in this clasp position the pair rolls over sideways, with the male on top. Both fish quiver intensely, give a sudden jerk and a few eggs and some sperm are released and fall onto the vegetation. The pair separates, the male often resumes courting, and the entire sequence may be repeated. Similar behaviour was observed for some of the same species in their natural habitat in Sri Lanka (Kortmulder et al., 1978), thus supporting the validity of the aquarium observations.

*Rasbora* is another Asian cyprinid genus in which pair-spawning is common. A well-known species is *R. heteromorpha*. A courting male either swims just in front of the female with exaggerated swimming undulations and all fins fully spread, or he "rides" above and slightly behind her. The close following of female by male occurs most often among thick vegetation and probably helps the fish to maintain visual contact (Wickler, 1955). Spawning occurs under a plant leaf. Both fish roll over through 180° and the male folds his tail over the female's back behind her dorsal fin in a spawning clasp (Fig. 5.2). The fish quiver and, while their genital openings are close together, a few eggs are extruded and stick to the underside of the leaf. The fish separate, right themselves, and the male may resume courtship.

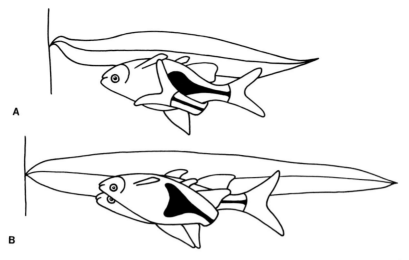

**Fig. 5.2.** Spawning pair of *Rasbora heteromorpha* seen from both sides. **A** female in front, **B** male in front. (After Wickler, 1955)

*Cyprinodontidae (Killifishes or Toothcarps)*. This large family has representatives in shallow fresh and brackish waters of all continents except Australia (Foster, 1967). Although the males of some species defend breeding territories after spawning (Echelle, 1973), the duration of territoriality is not known for most species, and direct parental care of the eggs does not generally occur (Kodric-Brown, 1977). The Florida flagfish (*Jordanella floridae*) is an exception, in that the male guards and fans the eggs throughout their development (Mertz and Barlow, 1966).

Observations of breeding behaviour in nature have been made of several species in the genera *Fundulus* and *Cyprinodon*. The same basic breeding pattern is found in *F. heteroclitus* (Newman, 1907), *F. diaphanus* (Richardson, 1939) and *F. notatus* (Carranza and Winn, 1954). Mature adults gather in shallow weedy areas, one or more males chase a female until she stops, then one male moves beside her, and pushes her against the vegetation. He maintains contact by folding his dorsal and anal fins over hers (Fig. 5.3A), or by hooking his dorsal, pelvic and anal fins underneath hers (Fig. 5.3B). The bodies of the two fish form a shallow S-bend, they quiver vigorously and spawn. In *F. notatus* the quivering bout ends with a vigorous, wide-amplitude tailbeat and extrusion of a single egg (Fig. 5.3A). In the other two species a few eggs are scattered onto the substrate during each of several rapidly repeated spawning bouts.

In *Cyprinodon* (the pupfishes) spawning occurs in shallow water, where adult males maintain territories. When a ripe female approaches, a male swims to meet her and performs a number of activities serving to keep her within the territory and ultimately to spawn. In most species one egg is laid with each spawning act, although a pair may perform several such acts in quick succession before separating (Kodric-Brown, 1977).

In *C. variegatus* the female is chased by the territorial male. She swims in circles, then dives and bites at the substrate and the male quickly moves to a position beside her. The two fish make lateral contact, with their bodies in a shallow sigmoid-bend,

96

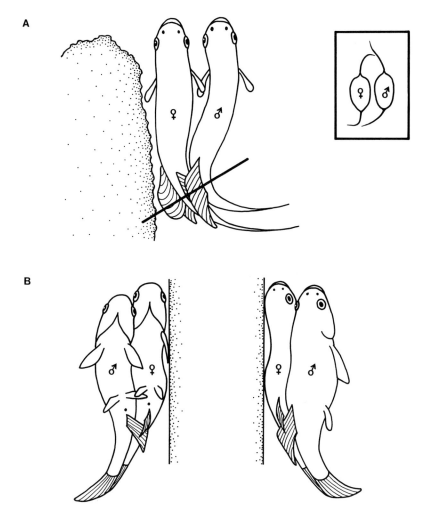

**Fig. 5.3.** Spawning in *Fundulus*. **A** *F. notatus.* Inset (*top right*) is cross-section at *heavy black line*, showing dorsal and anal fins diagrammatically. (After Carranza and Winn, 1954). **B** *F. heteroclitus*, from below (*left*) and above (*right*). (After Newman, 1907)

and the male clasps the female by folding his anal fin around hers so that their genital areas are close together. They jerk sharply, the female releasing one egg, and the male releasing sperm. This spawning act may be repeated four or five times in rapid succession (Raney et al., 1953; Itzkowitz, 1974a). Biting at the substrate by the female seems to be a visual signal to the male that she is ready to spawn. It does not appear to be a functional feeding act, since the female often spits out substrate materials after biting, and it is almost always followed immediately by the male joining her and the two fish clasping against the substrate. It has been described for *C. elegans* (Itzkowitz, 1969), *C. macularius* (Barlow, 1961) and *C. rubrofluviatilis* (Echelle, 1973).

Although courtship is generally brief and simple within this genus, it can be quite elaborate in *C. rubrofluviatilis*. A territorial male often *zig-zags* in front of one or several females swimming near. This is a series of sharp turns alternately to left and to right. He may also perform *headflicking*, a rapid sideways movement of the head, and *looping*, in which he swims back and forth in front of the female. Zig-zagging, head-flicking and looping are enticement acts by which the male stimulates the female to approach the spawning place (Echelle, 1973). The male may also perform *herding* and *steering*, as he appears to be trying to force the female to move towards the centre of his territory. In herding he stays behind and to one side of the female, keeping between her and the territory boundary. While steering he stays right beside her as she swims in an arc or a series of spirals, within the territory. Finally, the female nips at the substrate, and the male moves close beside her in the *sidling* act. Both fish then bend their bodies into a sigmoid shape and *clasping* occurs as he wraps his anal fin around hers. He presses her against the substrate, the two fish quiver and spawn with a sharp jerk movement. The male may clasp her again immediately and five or six spawning sequences can follow in rapid succession. More often, the spawning act is followed by another bout of courtship or by the female leaving the area (Echelle, 1973).

A lek type of breeding system was observed in a population of *Cyprinodon* sp. in a small lake with limited suitable spawning habitat (Kodric-Brown, 1977). Males competed for optimal spawning sites, and two types of male were successful: large, brightly coloured, territorial males, and smaller, non-territorial, "satellite" males. The success of the latter, despite their inability to establish territories, was apparently due to females preferring to breed with similar-sized males. Smaller females spawned with satellite males, larger females with larger, territorial males. Among the latter, the most brightly coloured had the greatest reproductive success, and the breeding habitat has therefore been termed a "lek" (Kodric-Brown, 1977).

*Comments.* Spawning of this type, where eggs and sperm are released onto unprepared substrate, occurs among small groups of several males with one female, or among pairs. Even in huge schools of herring, spawning does not seem to be simply a synchronized random release of gametes. Among the group-spawners there is little specialized courtship; males chase, push and butt the female and each other. Gamete release is probably synchronized by tactile and chemical signalling. In pair-spawners, male courtship consists mainly of leading the female towards a suitable site, followed by holding her in a spawning clasp. Visual locomotor displays by the male are more frequent early in courtship; tactile displays more frequent later, culminating in the direct clasping act which appears to trigger spawning.

Although there is no extended parental care, in most cases the eggs receive protection, either by being released close to the substrate, where they settle among gravel or dense vegetation, or by being placed directly onto leaves (often on their underside). In either case they are hidden from visually-hunting predators. Herring eggs, on the other hand, are exposed to several sources of mortality. The vast numbers of eggs in shallow water are vulnerable to predation, to exposure when washed onto beaches during storms, and to suffocation when spawning is intense and eggs are many layers thick (Outram and Humphreys, 1974).

The documentation of the "satellite" male phenomenon, and of differential reproductive success, among mature males of some cyprinodont populations, is strong evidence for the value of detailed observations in the field.

### 5.2.2.2 Eggs Buried in Substrate

This section concerns several groups of fishes that bury their eggs during spawning and then leave them, usually within a short time.

*Petromyzontidae.* Lampreys spawn in nests excavated in the gravel substrate of streams and rivers (see Chap.4). The same general pattern of spawning is followed in all species. *Petromyzon fluviatilis* of Europe and *P. marinus* of North America have been studied most closely (Applegate, 1950; Bahr, 1953; Hagelin and Steffner, 1958; Hagelin, 1959; Sterba, 1962; Hardisty and Potter, 1971).

While the male *P. fluviatilis* is digging a nest, the female repeatedly circles around him. Each time she passes the male she settles close above him so that her genital area appears to contact his head briefly. Finally she settles into the nest, attached to a stone by her oral disc. The male approaches from behind, contacts her body with his oral disc and slides forward to her head, where he attaches himself firmly (Fig. 5.4). He then wraps his tail tightly around the female, often just in front of her anterior dorsal fin. His tightly coiled body moves further backwards to the notch between her dorsal fins, at which point his genital papilla is close to her genital opening. This backwards movement of the male's tail may separate some eggs from the remainder, so that they are easily extruded during the spawning act (Hagelin, 1959). The female then shakes her tail and body violently and eggs and sperm are released together. These vigorous flapping movements stir up sand and gravel that settles and covers the eggs. In *P. marinus* the spawners separate, one or both anchor themselves at the upstream edge of the nest and with vigorous lashing movement of the body stir up substrate particles that bury the eggs. In both species the entire spawning act is repeated several times at intervals of from one to a few minutes, until all eggs have been expelled. The animals then leave the area and die within a few days.

One source of variation among lamprey species is the number of breeding adults together in a nest. In *Ichthyomyzon castaneus* about 50 adults were seen digging and spawning in a single large nest (Case, 1970). Spawning within the large group occurred in the basic single-pair system, with a male wrapped around an attached female. Occasionally a third lamprey attached itself to the male's head, and once five lampreys were seen attached this way in serial order. However in all cases only the original male was wrapped tightly around the female. *I. gagei* has been seen spawning in groups of 5 to 20 or more per nest, although aquarium observations of the same population showed a single-pair spawning system existed (Dendy and Scott, 1953). In *Lampetra japonica* and *L. richardsoni* spawning occurs in single pairs, although up to six individuals per nest have been seen in *L. japonica* (Heard, 1966; McIntyre, 1969). *L. planeri* in Europe spawns in groups of from 4 to 15 animals, within which, if there is a surplus of males, a female may have two males coiled around her (Sterba, 1962).

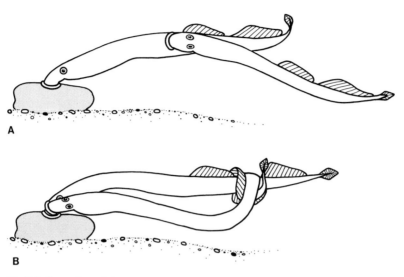

**Fig. 5.4.** Spawning of *Petromyzon fluviatilis*. **A** male sliding forward in contact with female, **B** male attached to female's head and with tail wrapped around her body. (After Hagelin and Steffner, 1958)

*Atherinidae (Silversides)*. Spawning by the California grunion (*Leuresthes tenuis*) is highly unusual, both in its location and the accuracy of its timing. The story has been told often, in the popular literature and on film, and was described in detail by Thompson and Thompson (1919) and Clark (1938).

At two-week intervals from March to August, during the few nights just following the full moon and the dark of the moon, adult grunion breed on the sandy beaches of southern California. The first fish are found on the sand as the waves recede shortly after the peak of the highest tide. Large numbers of fish come ashore, and ripe females quickly dig vertically, tail-first, into the soft, moist sand, to about the level of their pectoral fins. Up to five males may be lying flat on the sand near such a female; usually one curls around her so that their bodies are in contact. Eggs and sperm are shed at this point and the female then extricates herself from the sand by slowly bending from side to side. Both males and females then flop into the water and swim away from shore. The entire process of egg-laying takes about 30 s. It is not known how many of the males near each buried female shed sperm. Presumably any sperm released in the immediate vicinity of the female can pass down through the loosely packed, moist sand grains and contact the eggs.

The extraordinary feature of this breeding system is that the eggs are protected from predators by being buried several inches deep in the sand, and in a zone of the beach that will not be inundated until the next series of highest tides, two weeks later. When these tides arrive, the waves scour the shoreline and the young grunion, that hatch at about the same time, are washed out into the sea.

*Salmonidae (Salmon, Trout and Char)*. The general pattern of breeding events is similar in most salmonid species, both anadromous and freshwater. Mature males and females move onto the spawning grounds, which in most cases are shallow, gravel-bottomed areas of freshwater streams (Scott and Crossman, 1973). The

100

female selects a spawning location, excavates a nest in it (Chap. 4), and aggressively keeps other females away. One or more males stays near her and competes for status as her spawning partner. After each spawning bout the female covers the eggs with gravel. In *Oncorhynchus* the spawned-out adults die within a few days or weeks; in the other genera they move away and gradually recover. There is no extended parental care of the eggs, although some *Oncorhynchus* females may remain at the nest, keeping other fish away, for up to two or three weeks until they weaken and finally die (Briggs, 1953; Hanamura, 1966; Hanson and Smith, 1967; Heard, 1972).

The behavioural acts associated with spawning include digging, probing and covering by the female (see Chap. 4); quivering, crossing-over and a variety of body contact acts, such as pushing, butting and rubbing by the male; gaping and quivering by both sexes while spawning.

*Quivering* is a burst of high-frequency, low amplitude lateral undulations of the body, with the greatest amplitude at the head (Fabricius and Gustafson, 1954). The males of all salmonids quiver before spawning, usually when close to the female, hence she probably receives low frequency vibrations or direct tactile stimulation from the act. In most species the male is stationary while quivering. However, male *Salvelinus* quiver while gliding upstream past the female in the nest (Smith, 1941; Needham, 1961). Male *Salmo salar* often quiver and then snap at an inactive female in her nest. Also they may quiver near another approaching male, and attack if the intruder does not quickly leave. Thus quivering is used to stimulate females, but is also associated with aggression (Jones, 1959).

*Crossing-over* is a sideways movement by the male, back and forth across the female's caudal peduncle (Hartman, 1969). It has not been described as a pre-spawning behaviour for all salmonids, possibly because some observers did not consider it a regular part of courtship. In all species the attendant male contacts the female before spawning by nudging, pressing or pushing against her (Scott and Crossman, 1973). More careful analysis of these movements might show some regularity in the male's movements at the tail region of the female. Crossing-over has been described for *O. keta* (Tautz and Groot, 1975), *O. nerka* (McPhail and Lindsey, 1970), *S. gairdneri* (Hartman, 1969; Tautz and Groot, 1975), *S. clarki* (Smith, 1941), *S. fontinalis* (Needham, 1961), and *S. malma* (Needham and Vaughan, 1952). Hartman (1969) suggested that the crossing-over sequence by a courting male serves both sexual and aggressive functions because he can react quickly to any movements by the female, and he can also move quickly to interact with other males approaching from either side of the nest.

During the spawning act both male and female open their mouths maximally in a *gape*, and their bodies quiver violently. The female usually gapes first, and when she does the male darts into the nest beside her, suggesting that her gaping is a visual signal. The male in some species tilts sideways $10°$ to $15°$ from the vertical while spawning, with his ventral surface pressed close to that of the female (Mathisen, 1962). The gaping, quivering, and tilting are probably all involved, through visual and tactile signalling, in synchronizing the release of gametes.

Few quantitative data are available on temporal aspects of salmonid pre-spawning activities. Tautz and Groot (1975) provide some data for *O. keta* and *S. gairdneri* (Fig. 5.5). These show that while female digging bouts gradually declined

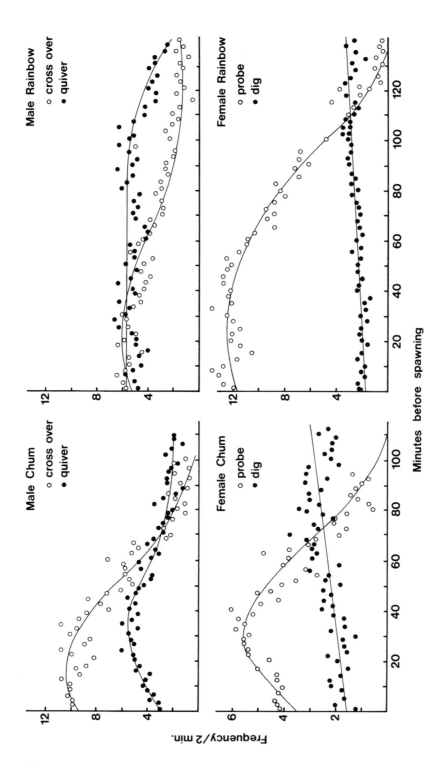

**Fig. 5.5.** Mean frequencies per 2 min of some prespawning reproductive behaviours by male and female chum salmon (*O. keta*) and rainbow trout (*S. gairdneri*). (Modified after Tautz and Groot, 1975)

in frequency during the 2 h before spawning, female probing acts rose to a peak about 30 min before spawning, then fell off in *O. keta* and remained steady in *S. gairdneri*. Quivering and crossing-over by males of both species increased in frequency as female probing bouts increased. This relationship was most pronounced in *O. keta* shortly before spawning, when a female probing act was almost always immediately followed by a male quivering bout. Thus, quivering, crossing-over and probing appear to be the critical synchronizing events of courtship.

One exception to the above description of salmonid spawning behaviour is the North American lake trout, *Salvelinus namaycush*. This species spawns in lakes (occasionally in rivers, Loftus (1958)) on rocky or rubble substrate at depths from 1 to 30 m (Royce, 1951; Martin, 1957). The males gather first in the spawning area and sweep the bottom clean of silt and debris by fanning with their tail and probing between the stones with their snout. When females arrive, the males court them by butting and occasionally by weaving back and forth underneath, apparently brushing the female's vent area with the back and dorsal fin (Martin, 1957). Spawning occurs among pairs or small groups; Royce (1951) saw seven males and three females in one tight group pressing together and quivering in unison while spawning. No nest is excavated, and after spawning the eggs are apparently not covered by the adults; they simply fall among the stones and are deserted.

All of these observations were made on lake-spawning *S. namaycush*; similar records are apparently not available for river-spawning fish. Detailed comparisons of spawning behaviour in the two habitats might provide valuable insights into the evolution of salmonid breeding behaviour patterns, especially the pre- and post-spawning substrate-directed behaviours. Likewise, it might prove useful to compare the breeding behaviour of those sockeye salmon (*Oncorhynchus nerka*) that spawn along lake shores with that of the vast majority that spawn in flowing streams.

*Salmonid Mating Systems*. In the three genera (*Oncorhynchus, Salmo* and *Salvelinus*) the basic mating system is promiscuity, with competition among males for access to spawning females. Females establish nesting territories to which one or more males are attracted. Aggressive competition among the males results in one becoming dominant; he is usually the largest. He attempts to drive the other males away, but one or more may stay nearby. Once an area is fully utilized by nesting females and their attendant males, additional females are attacked and driven off, usually by females (Foerster, 1968). When spawning begins in a nest, surplus males in the vicinity approach and may dart in and join the pair in the act of spawning. Presumably these males release sperm. A case of one female *O. gorbuscha* spawning simultaneously with six males is described by Wickett (1959). Usually a smaller number of accessory males, if any, intrude during the spawning act.

In some species the intruding accessory males are about the same size or slightly smaller than the dominant male; in others, they are much smaller. The latter include *O. kisutch, O. nerka*, and *O. tshawytscha*, in which some males ("jacks") return to freshwater on the spawning migration after having spent only about one year at sea. In *Salmo salar* some male "parr" (the freshwater juvenile stage) become sexually mature and spawn with large, sea-run fish. By cutting the sperm ducts of large

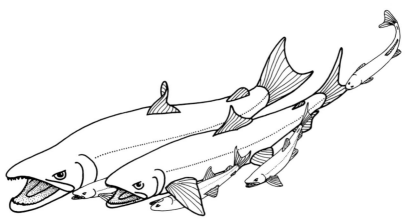

**Fig. 5.6.** Four male kokanee salmon entering nest of spawning pair of large *O. nerka.* (After McCart, 1970)

males held in a stream tank along with mature females and male parr, Jones and King (1952) showed that some of the parr were indeed sexually mature and released sperm as they darted into the nest beneath the spawning pair. Up to 75% of male parr in some British rivers become sexually mature during each breeding season (Orton et al., 1938), but the proportion of eggs fertilized by parr sperm is unknown. Female parr do not mature while in freshwater (Jones, 1959).

One study of spawning *O. nerka* showed that large males may defend a territory that includes more than one nest, each with its own female (Hanson and Smith, 1967). In an extreme case one male tried to hold six nesting females at once. He spent most of his time fighting rival males and very little in courting females; eventually he lost all but two of his females to other males. In another study from one to eight "satellite" male *O. nerka* held position on the downstream side of some nests occupied by a large pair. A few of these "satellites" were medium sized; some were "jacks" and most were kokanee (a small, non-migratory form of *O. nerka*, common in some parts of its range). Each occupied a distinct place near the nest and defended it against other "satellites". If the dominant male left the nest, the largest "satellite" male moved in and took his place with the female. As the pair approached spawning the "satellite" males became increasingly active and several darted into the nest during each spawning act (McCart, 1970). Figure 5.6 shows the relationship between a pair of large *O. nerka* and several satellite kokanee males at the moment of spawning.

*Percidae (Darters).* Some species of darters (North American percids) bury their eggs while spawning and then desert them. The logperch, *Percina caprodes,* provides one example (Winn, 1958a,b). In lakes mature males congregate over sandy shoals. Approaching females are followed by one or more males. When a female stops on the bottom she is mounted by a male, who grasps her with his pelvic fins and presses his tail down beside hers. The pair quiver vigorously and spawn, their movements resulting in the female being partly buried in the sand. They then separate, leaving the eggs well buried. One female may spawn repeatedly with different males.

104

The stream-living rainbow darter (*Etheostoma caeruleum*) spawns in much the same way (Reeves, 1907; Winn, 1958b). A ripe female moves from a pool onto a shallow stream riffle and is immediately approached by a male who vigorously chases other males away. She settles onto the bottom, and with vigorous lateral undulations digs herself part way into the gravel. Then the male mounts her and spawning occurs as the pair quivers vigorously. The fish then separate, leaving the eggs buried in the gravel, and each goes on to spawn again with other partners. In both *P. caprodes* and *E. caeruleum*, one or more additional males often dashes in beside the spawning pair and quivers as though trying to fertilize some of the eggs. Other darter species that leave their eggs buried after spawning in the same manner are *Etheostoma camurum* (Mount, 1959), *E. radiosum* (Scalet, 1973), and *Percina peltata* (New, 1966).

There is little if any preliminary courtship behaviour in these darters. This may be associated with the lack of a discrete, localized spawning site. Once a ripe female arrives on a shallow riffle or shoal, mature males attempt to mount her directly. There is competition for this favoured position, and thus the larger, heavier males are usually successful in mounting, but no elaborate exchange of signals is apparently needed to synchronize behaviour or stimulate the partners.

*Comments.* Fish that bury their eggs while spawning and then leave them, must select spawning sites that provide the developing eggs with two essentials. One is protection against predation; this is provided by a suitable depth of overlying sand or gravel. The other is a range of physical conditions (especially temperature, oxygen levels, waste removal) conducive to proper egg development. These are provided by the flowing water characteristic of the spawning grounds of most of these fishes. Species that spawn along lakeshores or sea coasts leave their eggs where they are subject to some sub-surface water movement.

The basic spawning unit is one male and one female, often within a breeding territory. There may be competition among males for access to a spawning female. This comes from neighbouring or satellite males (the latter usually much smaller than the paired male) who join the pair at the moment of egg release. This probably ensures fertilization of all eggs, which are quickly buried, either by the spawning act itself, or by special post-spawning burying movements.

One advantage of such a breeding system is that even large fish (some mature salmonids weigh 10–15 kg or more) can leave their fertilized eggs in small, shallow streams, where they are protected by the overlying gravel for an extended developmental period.

## 5.2.2.3 Eggs Placed Inside Shelters

A more specialized form of egg protection is shown by fishes that place their eggs inside natural, ready-made shelters, such as crevices in the substrate, or the shells of living molluscs and crustaceans. For example, the stream-spawning cyprinids *Notropis spilopterus* and *N. whipplii* spawn into crevices among stones and under loose bark respectively. In both species males actively defend small territories around their spawning sites, and often make passes over the site while quivering. This attracts females and spawning soon follows. The male *N. whipplii* settles onto

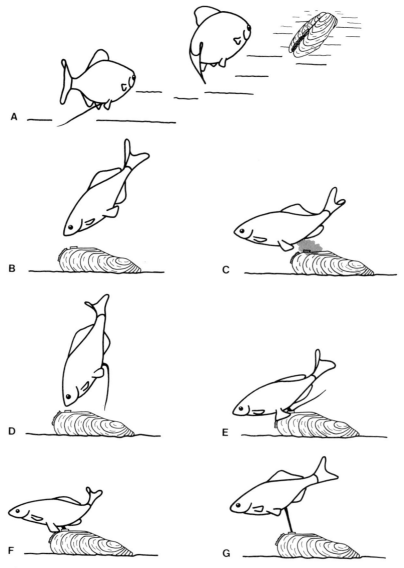

**Fig. 5.7A-G.** Spawning in *Rhodeus amarus*. See text for details. (After Wiepkema, 1961)

the back of the female, appears to press her into the space under a piece of loose bark, and both fish quiver as they spawn (Pflieger, 1965). The male *N. spilopterus* settles close beside the female, and they quiver while spawning into a crevice (Gale and Gale, 1977).

In an experimental study of spawning site selection, Gale and Gale (1976) showed that *N. spilopterus* spawned much more frequently into the crevices between vertically stacked black acrylic discs than between horizontally stacked discs, and most eggs (mean diameter 1.41 mm) were placed in narrow (1.5 and 3.0 mm) rather than wider spaces between discs. Apparently the fish selectively placed their adhesive eggs onto surfaces that received a steady water flow and were to some degree shielded from overhead light.

Fishes that place their eggs inside the shells of other living animals include some marine liparids (Cyclopteridae) and freshwater cyprinids. Among the former, virtually nothing is known about the behaviour of the spawning fishes. Large egg masses and larvae of the snailfish *Careproctus* have been found inside the gill chambers of the Pacific box crab *Lopholithodes foraminatus* (Parrish, 1972; Peden and Corbett, 1973), but spawning has not been observed.

The freshwater bitterling (*Rhodeus amarus*), that lays its eggs inside living mussels (genus *Unio* and others), has been studied in detail by Wiepkema (1961) and other workers (see Breder and Rosen, 1966). A mature male occupies a territory around a mussel, aggressively keeps other males away, and courts approaching females. He leads a female towards the mussel while quivering and holding his tail curved to one side (Fig. 5.7A). When she arrives at the mussel the male performs a head-down posture (Fig. 5.7B) and then skims over the mussel siphons (Fig. 5.7C). He may release sperm while skimming, even though the female has not yet spawned. He stays beside the mussel and quivers continuously. When the female is ready to spawn she holds position over the mussel in a steep, head-down posture, touches the exhalent siphon with the erect, thickened proximal part of her ovipositor, and then inserts the stiffened ovipositor into the siphon and releases from one to four eggs deep within the mussel's gill chamber (Figs. 5.7D, E, F). She then withdraws the now-slackened ovipositor (Fig. 5.7G), and the male repeatedly skims and releases sperm that is drawn into the mussel in the inhalent current. The sequence is repeated until the female's complement of eggs has been spawned, and both fish leave the area.

The main feature of spawning by fishes in this group is the careful, precise placement of eggs in a location where they are protected against predation and yet are bathed by moving water. The strategy reaches its most specialized form in the symbiotic use of space inside living invertebrates as a chamber for the developing eggs.

## 5.2.3 Species with Demersal Eggs and Prolonged Guarding

This group includes many species whose breeding behaviour has been studied intensively. Effort has been concentrated here because many of these fishes spawn readily in captivity, and some are also accessible for study in the field, where they breed in relatively clear, shallow water.

### 5.2.3.1 Eggs Released onto Substrate

In these species either the male alone or a female–male pair selects and defends a spawning site, and prepares it by cleaning it or excavating a shallow nest. After spawning one or both parents guard the adhesive eggs and the newly hatched young until the latter disperse.

*Centrarchidae.* Breeding behaviour in this family is best known among the sunfishes, genus *Lepomis* (Breder, 1936; Miller, 1963). In the spring and early summer, in response to increasing photoperiod and water temperature (Smith,

1970), mature males move into shallow water and establish territories in which they dig nests (see Chap. 4). Nests are clustered into colonies in some species, more widely spaced in others (Clark and Keenleyside, 1967; Keenleyside, 1978a). Females are non-territorial; when a mature female approaches nesting males, she is courted, and eventually enters a nest and spawns.

The basic courtship pattern of male *Lepomis* consists of swimming directly towards a female, displaying at her with his opercula spread open, occasionally butting or biting her, then returning quickly to his nest. In most species this sequence is performed in the horizontal plane, but the male *L. megalotis* usually rises as he approaches the female, tilts sideways above her, while quivering in bursts, then dashes back to his nest (Keenleyside, 1967). The quivering bouts probably coincide with sound production, which is typical of courting male *Lepomis* of several species (Gerald, 1971). On entering a nest the female settles to the bottom and turns slowly in a tight circle. The male circles with her, always on the outside. Periodically she rolls sideways, with her ventral surface pressed close to the male, quivers and releases some eggs that settle onto the substrate. They are fertilized by the male, who releases sperm while remaining upright. Spawning may continue until the female has released all her eggs or, more commonly, she moves on to spawn with other nearby males. The sequence is frequently interrupted, especially among colonial-nesting species, by the intrusion of other males at the moment of egg release, apparently trying to fertilize the eggs. Intruders include neighbouring territory-holders and non-territorial, satellite males who often gather around spawning pairs (Keenleyside, 1972b).

Promiscuity is the typical sunfish mating system. Each nest-guarding male usually spawns with several females during one or two days, then raises those clutches of eggs as a single brood. Females often deposit some eggs in each of several nests. When the young fish have dispersed from a nest the guarding male often breeds again, at the same or a different location.

*Pomacentridae.* In general the pattern of breeding behaviour in many pomacentrids is similar to that in sunfishes. Mature males establish territories on the substrate where they prepare a spawning site. The territories are often grouped in colonies. The males perform conspicuous displays towards females that remain nearby in schools. Eventually a female follows a courting male to his territory, spawning occurs and the male guards the eggs until they hatch and the young fish join the plankton.

The main activities of territorial males are aggression towards neighbours, preparing the spawning site, skimming and "signal jumping". In skimming the male swims slowly over the spawning site with his ventral surface brushing the substrate. "Signal jumping" is the name given by several authors to a conspicuous display sequence directed towards females. In it the male swims up from the substrate towards a nearby group of females, turns and quickly returns to his starting point. This is repeated a number of times until a female responds by following the male to the substrate. It has been described for several pomacentrid species, including *Chromis chromis* (Abel, 1961; Mapstone and Wood, 1975), *C. caeruleus* (Swerdloff, 1970; Sale, 1971a), *C. multilineata* (Myrberg et al., 1967; Albrecht, 1969), *Abudefduf saxatilis* (Albrecht, 1969; Fishelson, 1970), and

*Dascyllus aruanus* (Sale, 1970). There is much variation in the form of signal jumping, both within and between species. Probably these distinctive courtship sequences, together with characteristic male colour patterns, combine to produce species-specific signals that ensure intraspecific spawning.

When a female has followed a displaying male to his nest site, spawning occurs, either with the two fish close together, moving slowly over the substrate and quivering as they release eggs and sperm, or with the female first releasing eggs and the male following shortly after and fertilizing them. The female then returns to the school.

Among these substrate-spawning pomacentrids breeding is usually synchronized among the territorial males of one locality, and the basic mating system is promiscuity (Reese, 1964; Russell, 1971). Pairs are together only for the brief period of spawning and each male is likely to have eggs from several different females together in his nest.

*Cichlidae.* The mating system of cichlids that guard their eggs on the substrate (i.e., substrate-brooders) is based on the monogamous pair. Both members of a pair take part in all phases of the breeding cycle. After preparation of the spawning site the two fish interact with each other in a series of mutually stimulating displays and substrate-directed actions, culminating in the spawning act. This sequence is also found among cichlids that spawn in enclosed shelters (Sect.5.2.3.2), and represents the breeding strategy of all substrate-brooding as opposed to mouth-brooding cichlids (Sect.5.2.4). The distinction I am making between spawners on open substrate and spawners in enclosed shelters is for convenience only, since the available information on breeding behaviour of most cichlids comes from captive animals in aquaria, where potential spawning sites are determined by the observer. The work of Barlow and others in the lakes of Central America shows that many of the cichlids found there (some of which will spawn readily in exposed locations in aquaria) usually spawn in caves in the rocky substrate (e.g., Barlow, 1976; McKaye, 1977).

During establishment of the pair bond a male and female interact with a wide range of postures, displays, and movements (Baerends and Baerends-van Roon, 1950). Some of these are aggressive elements, and are also used by each member of the pair to keep intruders out of their territory. Others are more clearly sexual, judging by their increasing frequency as spawning approaches. Still others are directed at the spawning site, although they may have sexual as well as cleaning functions. In most species, all the actions are performed by both members of a bonded pair, although there are often slight but consistent differences between sexes in the frequencies of some acts (e.g., Greenberg et al., 1965; Chien and Salmon, 1972).

Courtship culminates in spawning at the prepared site. The female begins by releasing a series of adhesive eggs, one at a time, through her large, blunt ovipositor as she drags it slowly over the site. The male releases sperm as he, in turn, passes over the eggs. The two acts may occur in alternate bouts; however, it is not uncommon for a female to lay several strings of eggs before moving away to allow the male to fertilize the entire mass. On occasion both fish spawn simultaneously while moving over the substrate close to each other.

*Comments.* The species included in this section spawn in shallow water, on the substrate, in areas prepared by excavating in soft bottom or by cleaning off small areas of firm substrate. Either the male alone (centrarchids and pomacentrids) or male and female pairs (cichlids) maintain a territory around the spawning and brood-rearing site. These are often clustered into colonies within which breeding tends to be synchronized. Satellite or neighbouring males may intrude during spawning and attempt to "steal" fertilizations from the resident male.

The intensive care and protection provided for the progeny by one or both parents means that these fish are not restricted to areas of flowing water or dense cover for spawning. Colonial nesting tends to be associated with open-water areas, but there are few data available to suggest a causal relationship here. There is a need for quantitative studies on the adaptive significance of group versus solitary nesting in fishes.

### 5.2.3.2 Eggs Placed in Enclosed Shelter

This section deals with the breeding behaviour of several groups of fishes that excavate spawning and brood-rearing sites inside enclosures on the substrate.

*Pomacentridae.* Damselfishes that spawn in enclosed shelters have been studied in their natural habitat at widely scattered locations. Spawning sites used include empty mollusc shells, holes among rocks, and inside discarded automobile tires, pipes, and other refuse. In all cases males excavate and maintain the spawning site. The basic social system during breeding takes two different forms. One is similar to that of open substrate-spawning pomacentrids (Sect. 5.2.3.1); males occupy territories around spawning sites while females and sub-adults remain nearby in loose feeding aggregations. An example is *Pomacentrus nagasakiensis*, studied in Japan by Moyer (1975). In the other system both adult males and females maintain long-term territories on the bottom. Nest-occupying males have small territories, restricted to the area around the nest, which they defend with great vigour, especially when guarding eggs. Females and non-nesting males have larger, less actively defended territories in the same general area as the nesting males. Examples are *Eupomacentrus partitus, Abudefduf zonatus*, and *A. luridus*. These were studied respectively near the Bahama Islands (Myrberg, 1972), at Heron Island, Great Barrier Reef (Keenleyside, 1972a), and at the Azores (Mapstone and Wood, 1975).

In all four of the above species, breeding begins when a nest-occupying male moves out towards a female, and tries to attract her to his nest. In *A. zonatus* and *A. luridus* the male swims horizontally towards a nearby female, turns this way and that when near her, then swims straight to his nest, usually enters it, and makes several skimming passes over its inner surface. If the female follows, the male becomes more excited and, in *A. zonatus*, he quivers and nips gently at her head. When spawning begins the two fish alternate, the female laying a batch of eggs, then the male passing over them releasing sperm. The males of the other two species attract females by courtship movements in a vertical plane. The *P. nagasakiensis* male performs "signal jumps", in which he swims slowly and vertically up from the substrate, then turns and dives down. When a female approaches he performs "enticement" behaviour, similar to the courtship described above for the *Abudefduf*

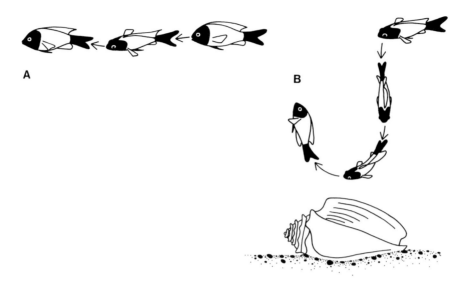

Fig. 5.8. Two stages of courtship by male *Eupomacentrus partitus*. **A** tilt sequence, **B** tilt and dip sequence above empty conch (*Strombus*) shell. (After Myrberg, 1972)

species. He swims about her while quivering, turning and twisting, and then dashes to his nest. An *E. partitus* male usually begins courtship while swimming some distance above the nest site. He tilts sideways, and may follow this immediately with a "dip" sequence in which he dives, turns sharply near the substrate and swims rapidly up again (Fig. 5.8). This is, in effect, the reverse of the up-down signal jumps of other pomacentrids. The male *E. partitus* also uses a direct approach to the female, nudging her ventral area, leading her towards his nest and, if she follows, orienting vertically head-down and quivering rapidly. The two fish then swim about close to each other in tight circles, the male enters the nest, skims over the inner surface, and spawning follows directly (Myrberg, 1972).

Some pomacentrids vocalize during courtship. For example, *E. partitus* makes a characteristic "chirp" while dipping, and "grunt" during tight circling at the nest. These and other sounds produced by this species have been intensively studied under natural conditions and in the laboratory (Myrberg, 1972; Myrberg and Spires, 1972).

An interesting activity that occurs occasionally among breeding pomacentrids is the close aggregating of several adults for a short period of intense, social interaction with each other. The fish leave their territories, congregate in a close group, within which there is mutual displaying, circling, parallel swimming, and close following, but little overt aggression. Courtship behaviour by males may occur in the clusters, which break up after a few minutes, with all participants returning to their territories. This has been described for *A. zonatus* (Keenleyside, 1972a), *P. nagasakiensis* (Moyer, 1975), and *Amphiprion clarkii* (Moyer and Sawyers, 1973). Its function is not certain, but it is clearly related in some way to reproduction and possibly, through mutual stimulation of near-neighbours, promotes the synchrony in spawning that occurs among the fish of one locality

111

(Keenleyside, 1972a). It may also promote sex recognition, especially in species that are sexually monomorphic (Clarke, 1971).

*Blenniidae*. These are benthic fishes that spend most of their time in burrows or under rocks in shallow water (Gibson, 1969). Sexually mature males defend territories around their shelters. When a ripe female swims near, a male begins courtship, by making conspicuous head movements, followed by swimming out and around the female trying to stimulate her to enter his enclosure. Specific colour changes, especially of the head, and the erection of head tassels in some species, increase the conspicuousness of male courtship. Details vary among species, several of which have been studied closely in nature.

For example, the courtship repertoire of male *Blennius canevae* includes rapid lateral head-shaking, which emphasizes the conspicuous black and yellow pattern of his head, zig-zag jump-swimming around the female, gentle butting, and leading, by darting into his hole and out again. When the female enters the nest-hole he often stays just outside and performs rapid head-nodding or "hammering". In *B. inaequalis* the courtship repertoire is simpler; the main differences are: the sexes are nearly monomorphic; both sexes perform more head-up postures towards each other, the female apparently displaying her swollen belly as she rears up; and the male shows less frequent zig-zagging, butting, and hammering (Abel, 1964). Courtship in *B. rouxi* is similar (Heymer and Auger de Ferret, 1976). The male is very conspicuous, with a dark horizontal stripe on a light yellow body, and two long head filaments. He rests in his shelter with head protruding and performs bursts of rapid head-nodding, often followed by a "signal jump" in which he dashes straight out 30 to 50 cm from his hole with all fins spread, then darts back into the hole again. He also performs up and down swimming movements while oriented vertically head-up in front of the female, and "shows the nest hole" to her by holding position with his head oriented towards the hole and vigorously head-nodding. The male *Istiblennius zebra* has an elaborate series of courtship acts including head-bobbing, vertical looping in front of his shelter, and dip-swimming while returning from a female. He may also swim repeatedly around the opening of the hole (Phillips, 1977).

In all these species the male either enters the nest hole with the female and they spawn together, or, if the hole is small, he waits outside until she emerges, then darts in to fertilize the eggs. The female is chased away after spawning; the male alone cares for the developing eggs.

*Cottidae*. Sculpins spawn in pits excavated under stones by the male. The courtship and spawning of two freshwater species of this large family have been described in detail. In *Cottus gobio*, the territorial male lies inside his nest looking out of the entrance. When another sculpin approaches he rushes out to challenge it, his head darkens, all fins are fully spread and he head-nods vigorously, together with loud knocking sounds. He usually bites at the other fish; if it is a ripe female he may engulf the entire fore part of her head in his mouth (Fig. 5.9). On releasing his grip, a female that is ready to spawn enters the nest, turns upside down and examines the roof onto which she eventually lays her adhesive eggs. Meanwhile the male moves about her, sometimes blocking the mouth of the nest by lying across the entrance,

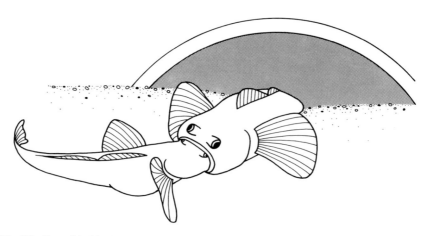

**Fig. 5.9.** Courtship bite by male *Cottus gobio* at entrance to his nest. (After Morris, 1954)

sometimes turning upside down in contact with her. After spawning, he chases her away and guards the eggs. More than one female may lay eggs successively in the same nest (Morris, 1954).

In *Cottus bairdi*, the sequence of events is similar, with the aggressive, territorial male rushing out towards an approaching female with head shaking and nodding (in the horizontal and vertical planes respectively), body quivering, and much biting. He bites at any part of her head, body, and tail, and may also engulf her head in his mouth. Eventually the female enters his nest and eggs are laid on the roof (Savage, 1963).

*Comments.* The shelter-spawners discussed here make use of relatively small enclosures on the substrate to protect their eggs. The shelter is guarded and kept free of drifting silt and sand by the male only; the females play no part in parental care. The adults are vulnerable to predation mainly during the relatively brief courtship periods, when the males are conspicuous as they stimulate ripe females to enter their enclosures. Otherwise all adults are either closely associated with the substrate, and its familiar escape routes, or are in schools that also provide protection.

Thus, this is an efficient breeding system for small, substrate-oriented fishes that are able to make use of small spaces in the complex substrate of coral reefs, rocky shorelines, and gravel-bottomed streams.

### 5.2.3.3 Eggs Placed in Constructed Nest

The best-known fishes in this category are the sticklebacks and the bubble nest-builders.

*Gasterosteidae.* The amount of published information on stickleback breeding behaviour ranges from a huge literature on *Gasterosteus aculeatus* to almost none on *Spinachia spinachia* (Wootton, 1976). A brief outline of courtship and spawning in the six best-known species is given below; details are compared in Table 5.1.

113

**Table 5.1.** Courtship and spawning behaviour of six stickleback species

| Behaviour Pattern | Initial approach to ♀ | ♀ Response | ♂ Leads towards nest | ♀ Response | ♂ Shows nest entrance | ♀ Response | ♂ Tactile stimulation | ♀ Response | ♂ Response | Authority |
|---|---|---|---|---|---|---|---|---|---|---|
| *Gasterosteus aculeatus* | Zig-zag dance, horizontal | Head-up, turns towards ♂ | Straight to nest | Follows ♂ to nest | Dorsal roll, jerky snout insertions | Enters nest | Quivers with snout against tail | Spawns and leaves | Enters nest, spawns and guards | ter Pelkwijk and Tinbergen, 1937; Wootton, 1976 (all species) |
| *G. wheatlandi* | Zig-zag dance, head-down | ditto | Head-down, body arched, quivering | Follows, snout close to his pelvic spines | Jerky snout insertions | ditto | ditto | ditto | ditto | McInerney, 1969 |
| *P. pungitius* a) European | Jump dance vertical, head-down | ditto | Jump dance to nest | ditto | Head-down fanning | ditto | ditto | ditto | ditto | Morris, 1958 |
| b) Lake Huron | ditto | ditto | Straight to nest | ditto | Head-down, no fanning | ditto | ditto | ditto | ditto | McKenzie and Keenleyside, 1970 |
| *Culaea inconstans* | Dart and pummel | Settles to bottom, head up | Body arched, exaggerated tail-beats | ditto | Horizontal fanning | ditto | ditto | ditto | ditto | McKenzie, 1969 |
| *Apeltes quadracus* | Dart and pummel | Settles to bottom, head-up | Swims on spiral path to nest | ditto | Head-down, tail-beats | ditto | Gentle biting of tail | ditto | ditto | Rowland, 1974a |
| *Spinachia spinachia* | Bite | Turns towards ♂ | Straight to nest | Follows ♂ to nest | Head jerks | ditto | ditto | ditto | ditto | Sevenster, 1951 |

Sustained courtship behaviour usually begins when the male has completed building a nest in his territory. As a ripe female approaches, he swims out to meet her. The form of the initial male approach varies among species. Both *Gasterosteus* species perform the well-known "zig-zag dance", which is a series of short, vigorous jumps, each along a slightly curved path and with the male alternately facing towards the female and facing slightly away from her on successive jumps (ter Pelkwijk and Tinbergen, 1937; Tinbergen, 1951). *G. wheatlandi* males orient about 30° head-down while zig-zagging (McInerney, 1969). In *Pungitius* the courting male approaches a female in a head-down, almost vertical posture and makes a series of quick, rhythmic jumps, turning slightly sideways with each jump (Morris, 1958).

The *Culaea* male approaches slowly at first and then lunges at the female and strikes her head region with his mouth in an act called "pummelling" (McKenzie, 1969). In *Apeltes* the male also moves slowly at first, then darts at the female and pummels her several times in rapid succession (Rowland, 1974a). The *Spinachia* male is even more direct, attacking the female and biting at her tail and fins (Sevenster, 1951).

In all species a responsive female reacts to this initial male approach by turning to face him. *Gasterosteus* and *Pungitius* females also pitch into a slightly head-up posture, in which their swollen abdomens are conspicuous. *Culaea* and *Apeltes* females appear to be driven into the substrate by the aggressive pummelling of their males; nevertheless they turn to face him and orient slightly head-up.

The next phase in courtship is called *leading*. The male swims towards his nest and a receptive female follows. Male *G. aculeatus* and *Spinachia* swim directly and quickly to the nest. The *G. wheatlandi* male pitches slightly head-down, arches the body and quivers rapidly as he moves slowly towards the nest. *Culaea* males also arch the body and swim slowly, with the tail beating in a rapid, exaggerated, wide-amplitude manner. The male *Apeltes* darts away from the female along a horizontal spiral path, which may cover from 90° to 360°. He stops, and if the female approaches, he swims towards the nest in a slightly head-down posture; otherwise, he continues spiralling. *Pungitius* males show some variation in leading. In the European form the male leads straight towards the nest while dancing in a series of head-down jumps, similar in form to the initial approach jumps (Morris, 1958). In a North American population studied in Lake Huron the male swims directly back to the nest from the female without dancing (McKenzie and Keenleyside, 1970).

A responsive female reacts to male leading by following him towards the nest. Often she stays close behind and below him, with her snout near his large ventral spines, which are held stiffly open during leading. Sometimes she nudges his ventral surface with her snout.

When the pair reaches the nest *Gasterosteus* males perform "showing the nest entrance" behaviour, by orienting with the head close to the entrance, and moving forward and back with short, jerky motions. *G. aculeatus* males also roll over about 90°, with the dorsal surface directed towards the female. *Culaea* males grasp the upper rim of the nest with the mouth and beat the tail in parental fanning fashion. *Pungitius* males orient head-down, and in this position the European *Pungitius* males show typical parental fanning behaviour. The Lake Huron fish do not; they simply hold position there by slight fin movements. *Apeltes* males orient head-down with the snout over the nest, give several vigorous tail beats, and then back away a short distance. *Spinachia* males perform head jerks at the nest.

In all species, this "showing" behaviour is a strong stimulus to the female to enter the nest. When she has entered it, the male provides tactile stimulation, in response to which she lays a clutch of eggs. In *Apeltes* and *Spinachia*, this takes the form of gentle biting of the female's caudal peduncle; in the other species the male presses his snout against her peduncle and quivers vigorously. When a clutch is laid, the female leaves the nest, the male immediately enters, quickly fertilizes the eggs, and then is highly aggressive, chasing all other fish, including the female, out of his territory.

*"Dorsal pricking"* is a conspicuous behaviour that may occur during courtship in *G. aculeatus*; it has not been described for other sticklebacks. The male fully erects his dorsal spines and manoeuvers to a position below the female. He then moves upwards as though trying to prick her ventrum with his large spines. The male may bite the female between bouts of dorsal pricking, especially if she remains near his nest (Wootton, 1976). An experimental investigation showed that dorsal pricking is more likely to occur when the relationship between the underlying motivational states of aggression and sex in a courting male is weighted towards aggression, and it declines in frequency as that balance shifts towards sex (Wilz, 1970). It serves to prevent a female from entering the territory of a male who is not yet ready to spawn, and is especially common in situations where males can be shown to be highly aggressive (Wilz, 1973). Observations of *G. aculeatus* breeding in nature showed that dorsal pricking may be a common occurrence among free-living fish, where its temporal occurrence in relation to aggressive behaviour supports Wilz's hypothesis (Wootton, 1972).

The summary in Table 5.1 shows that extensive variation in reproductive behaviour exists within this small family. Although much work has been done on morphological variations, especially in the widely distributed genus *Gasterosteus* (Wootton, 1976), the radiation in behaviour patterns remains largely unexplored. Some work on *Pungitius* (Morris, 1958; McKenzie and Keenleyside, 1970; Foster, 1977) suggests that detailed ethological comparisons, both in the laboratory and in nature, of fish from different habitats would provide useful insights into the evolution of variation in stickleback reproductive patterns.

*Belontiidae (Bubble Nest-Builders)*. Most of these fishes breed in the quist, shallow waters of swamps, marshes, and inundated rice fields of southeast Asia (Forselius, 1957). Virtually all the information on their reproductive behaviour has come from aquarium studies. The most characteristic feature is the floating nest, made of small air bubbles that are produced and packed together by the male, who actively defends a territory around his nest (see Chap. 4). An elaborate courtship between a single male and female culminates in spawning just below the nest; the fertilized eggs are retrieved by the male and placed in the nest.

Although the species differ in some details, the overall pattern of courtship and spawning is similar among many species in the genera *Colisa, Trichogaster* and *Macropodus* (Forselius, 1957; Miller, 1964; Hall, 1968; Hall and Miller, 1968; Machemer, 1970; Miller and Robison, 1974; Robison, 1975; Vierke, 1975). The initial reaction of a nest-guarding male to an approaching conspecific female

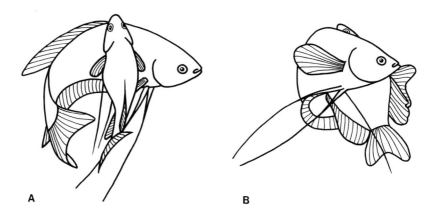

**Fig. 5.10.** Curving (**A**) and clasping (**B**) by male *Trichogaster trichopterus*. Female is head-up in both cases. (After Miller, 1964)

depends largely on her sexual motivation. If she is ready to spawn she may immediately move towards the male, butt against his side with her snout, and swim to the nest. If she is less strongly motivated the male approaches and tries to lead her to the nest, usually with median fins fully spread. At the nest, the female orients towards the male's side and tilts head-up; he begins to encircle her with his body by curving his head and tail towards her. These movements continue until the male clasps the female tightly within the curve of his body. Curving and clasping in *Trichogaster trichopterus* are shown in Figure 5.10. The female rolls over so her ventral surface is oriented upwards, and then with their genital pores close together both fish quiver violently as eggs and sperm are released. At this point, the two fish remain immobile, and drift slowly apart for several seconds as the fertilized eggs either sink or float, depending on the species. This brief post-spawning state of immobility has been called "swimming inhibition" (Forselius, 1957). It ends abruptly as the male chases the female away, and retrieves the eggs. He picks up several in his mouth, moves them about with chewing movements, and spits them into the bubble nest. While in his mouth the eggs become coated with a mucous secretion by which they adhere to the nest (Forselius, 1957). Several such spawning bouts usually occur during a breeding cycle by one pair.

*Comments.* Both sticklebacks and bubble nest-builders make elaborate receptacles for protecting their eggs. Sticklebacks place their nests in a wide variety of locations; the only essential requirements seems to be that the adhesive eggs must be maintained at one locus, where the male can guard, aerate, and clean them. Belontiid bubble nests float at the surface in quiet, shallow water. In both groups the female must be induced to lay her eggs in or near the nest, thus allowing the male to maintain close guard over the structure in which he has so heavily invested. Males often have clutches from several females in one nest, and females spawn with different males as successive batches of eggs mature.

### 5.2.4 Species that Carry Their Eggs

Several categories of egg-carrying fishes are recognized by Balon (1975). These include species that carry batches of fertilized eggs attached to a hook on the forehead (*Kurtus*); eggs attached singly to the ventral body surface by vascularized, cup-like extensions of the skin (*Aspredo* and *Bunocephalus*); eggs enveloped in a ventral pouch (*Loricaria vetula*, *Hippocampus* and *Syngnathus*); and finally eggs carried inside the mouth of one or both parents. Little is known about courtship and breeding in any of these groups except for the oral brooders, and here the cichlids are the best-known representatives.

*Cichlidae.* Mouthbrooding the eggs is widespread among African cichlids (Fryer and Iles, 1972) and also occurs in the South American genera *Aequidens* and *Geophagus* (Timms and Keenleyside, 1975). There is an extensive literature on the subject, and the breeding cycle of the South African cichlid *Hemihaplochromis philander* will serve as an example of the strategy.

The mature male *H. philander* establishes a territory in which he digs a depression. Courtship begins when a ripe female approaches him. Ribbink (1971) recognized eight stages in the courtship and spawning sequence:
(1) The male quickly approaches the female, performs a lateral display in front of her, with all fins spread and branchiostegal membranes lowered; he then performs "side-shake", in which all his fins quiver vigorously. (2) If the female holds position the male faces away, tilts slightly head-down and beats his tail vigorously in the "follow-shake" pattern. (3) This is quickly followed by "lead-swim"; the male swims towards his nest with exaggerated tail-beats and slightly head-down. (4) He then settles into his nest and quivers in "horizontal nest-shake". (5) If the female enters the nest, the male performs "vertical nest-shake", oriented head-up and with his tail pressed onto the substrate. Body and fin quivering is intense. The female approaches and gently butts his abdomen. (6) He in turn butts her abdomen, and they both circle in the nest, gently butting each other. (7) The female soon lays a small batch of eggs, while the male watches. When she stops laying, he moves over the eggs releasing sperm. She picks up all the eggs in her mouth, beginning while the male is still fertilizing them; thus she takes both sperm and eggs into her mouth. (8) When the eggs are picked up, the male performs "vertical nest-shake", with his anal fin spread open on the substrate. In this position the conspicuous red and black markings on the anal fin, and the contrasting black and white pattern on the pelvic fins, seem to provide "guide lines" towards the male's genital pore. The female nibbles at his genital region, presumably taking more sperm into her mouth. The spawning procedure in the nest is repeated until the female has laid and picked up all her eggs. She is then driven away by the male, who plays no further parental role.

Many species of maternal mouthbrooding cichlids have anal fin markings that are especially conspicuous in breeding males. The variation in form, size and colour of these markings, and their role in stimulating and directing the fin-nibbling movements of females during spawning, have been extensively studied by Wickler (1962a,b). He argued that they are in effect "egg-dummies" which the female mistakes for her own eggs, hence snaps at them, inadvertently stimulating the male to release sperm which is quickly taken into her mouth. In *Haplochromis burtoni*,

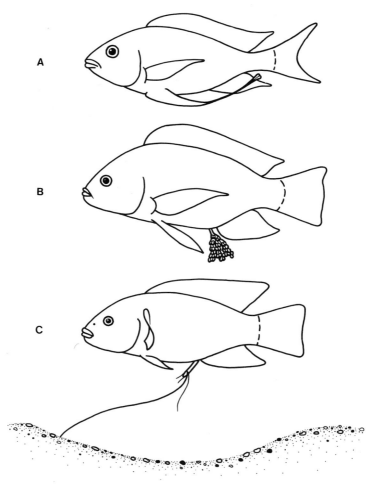

**Fig. 5.11.** Male cichlids with different devices used to promote fertilization in the female's mouth. **A** *Ophthalmochromis ventralis,* **B** *Sarotherodon variabilis,* **C** *S. macrochir.* (After Fryer and Iles, 1972 and Wickler, 1965a)

and probably many other species of this large African genus, the male spreads open his brightly spotted anal fin on the substrate in front of her as she picks up a batch of eggs. The fin quivers violently during this display, which is the equivalent of Ribbink's "vertical nest-shake". If she turns away, he shifts around, trying to keep his opened anal fin in front of her.

The males of some maternal mouthbrooding cichlids have other morphological devices that appear designed to induce the female to pick up her own eggs promptly (Fig. 5.11). *Ophthalmochromis ventralis* has brightly coloured, swollen tips (said to resemble eggs) on the elongated pelvic fin rays. *Sarotherodon karomo, S. variabilis,* and *S. rukwaensis* have a conspicuous tassel that develops behind the genital pore during breeding periods (Lowe-McConnell, 1956; Fryer and Iles, 1972). In *S. variabilis* males over 20 cm long, the tassel may be 6 cm in length, "...and consists of a mass of soft, pliable tissue drawn out into numerous bright orange lobules which, set against the white ground colour of the organ, renders the tassel very conspicuous" (Fryer and Iles, 1972, p. 111). The male *S. macrochir* not only has a genital tassel, but releases a long, white sperm thread, that the female picks up in her mouth before sucking up her eggs (Fig. 5.11C). This appears to be a form of spermatophore that prevents the sperm from dissipating quickly in a cloud (Wickler, 1965a, 1966b).

Among uniparental mouthbrooding cichlids, territorial males breed successively with different females. The females may have separate egg batches fertilized by one or more males, depending largely on the density of fish in the breeding area. If a pair is not seriously disturbed while spawning, the female's eggs will likely all be fertilized by one male. However, on an *S. macrochir* spawning ground, for example, where many nests were crowded close together, and the density of adults and juveniles was high, spawning pairs were frequently disturbed by egg-predators, and females spent only about one minute in each nest, i.e., long enough to have one batch of eggs fertilized (Ruwet, 1962, 1963). Likewise, a field study of *H. burtoni* showed that males nesting in crowded colonies were frequently disturbed during courtship by groups of fish containing non-breeding conspecifics and other cichlids (Fernald and Hirata, 1977).

## 5.3  Mating with Internal Fertilization

Most information on the breeding behaviour of internally fertilizing fishes comes from the ovoviviparous cyprinodontoid family Poeciliidae. In this group the anal fin of the male is modified into a slender, club-shaped organ, the gonopodium, through which spermatophores are transferred into the female's genital aperture. The spermatophore is encased in a viscous fluid, and during storage in the female, ovarian secretions lower the viscosity of this fluid, thus releasing sperms. Batches of mature ova can be fertilized at intervals by sperms from a single spermatophore (Rosen and Gordon, 1953). Females often show superfetation, i.e., two or more broods of young of different ages can develop in the ovary simultaneously (Turner, 1937). The gonopodium is highly variable in size, shape, and in the number and form of terminal hooks and spines that may serve several functions: as sensory receptors, organs of contact and insertion, and as holdfast devices for maintaining contact between the two fish and for steadying the entire structure during copulation (Rosen and Gordon, 1953).

Breeding in many poeciliids is preceded by elaborate courtship in which the relatively active and conspicuous male tries to induce the female to accept his copulation attempts (Fig. 5.12). The guppy *Poecilia reticulata* will serve as an example, because it is so well known (Baerends et al., 1955), and because of the rich and varied repertoire of male display patterns, many of which expose to the female the patches of bright colours and black patterns on his body and fins. Those occurring at early stages in courtship include: *following* the female, with median fins open or closed; *holding position* in front of the female; *biting* at her genital area; *"luring"*, in which the male folds all fins and, while facing the female's head, moves slightly forward and back, as though trying to induce her to move towards him (Fig. 5.12C). All of these actions seem to have the primary function of making

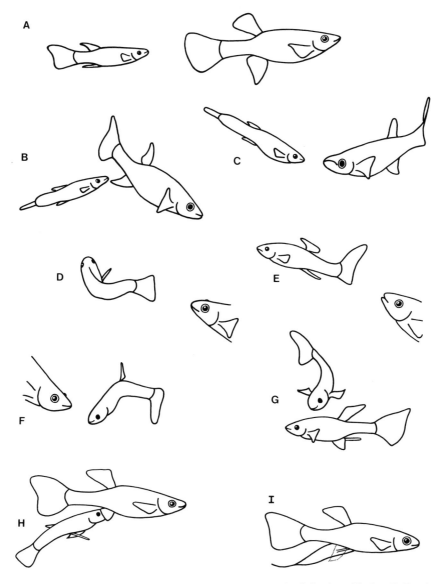

**Fig. 5.12A-I.** Some courtship patterns by male *Poecilia reticulata*. **A** and **B** following, **C** luring, **D, E** and **F** sigmoid displays, **G** display posture, **H** and **I** copulation attempt. (Modified after Baerends et al., 1955)

the female slow down or stop swimming. If she does, the male moves on to a series of *"sigmoid displays"*, so-called because his body and tail are bent into an S-curve, while he stays in front of her. At any stage during courtship, but especially during the performance of sigmoid displays, the male may perform *"copulation attempt"*. Here he swims quickly around behind the female, approaches her from slightly below, pitches head-up and tilts slightly in a ventral roll as he swings his gonopodium towards her genital area. His body forms an S-bend as he tries to insert the tip of his gonopodium into her genital pore (Fig. 5.12H,I). While the gonopodium is swung forward, one of the pelvic fins is also moved forward, and the gonopodium braced against it (Clark and Aronson, 1951; Rosen and Tucker, 1961).

In general, female *P. reticulata* are less active and conspicuous than males during courtship, yet they are not entirely passive. When aroused by male courtship, a female may perform: *gliding*, which is a slow forward progression, using only slight caudal and pectoral fin movements for propulsion; *arching*, a stationary posture with head and tail slightly elevated; *wheeling*, swimming in a tight circle, with the male beside and below her, while she is slightly arched, thus exposing her genital area; *copulation*, in which she accepts the male's gonopodial thrusts at her genital pore; and *wobble*, a jerky, uneven swimming with wide amplitude lateral movements of the body often performed immediately after copulation.

The importance of active cooperation by the female was well demonstrated in a comparative study of reproductive behaviour in four *Poecilia* species that occur sympatrically in Guyana (Liley, 1966). Selective responsiveness by females to the courtship behaviour of conspecific males was an important aspect of premating reproductive isolation among these species.

There appears to be a relationship between the complexity of courtship behaviour and the length of the male's gonopodium among poeciliid fishes (Rosen and Tucker, 1961). A comparison of species from 13 genera showed that those with relatively long gonopodia (e.g., *Pseudoxiphophorus* and *Girardinus*, in which the tip reaches to or beyond the male's eye when swung forward) lack conspicuous visual displays and physical contact during courtship. Those with relatively shorter gonopodia (e.g., *Poecilia* and *Xiphophorus*) have elaborate and conspicuous displays and do make physical contact before copulation. Rosen and Tucker (1961) argue that a fish with a very long gonopodium can see his own organ when it is swung forward, and can thus use visual cues to orient towards the female, even if she is active. One with a shorter gonopodium cannot see it, and therefore has a greater need for female cooperation. Hence the elaborate courtship display repertoire of these males is designed to induce the female to remain stationary, so that his copulation attempts have a greater probability of success.

Males of the genus *Gambusia* may use both visual cues from the female and tactile support of the gonopodium with a pectoral fin during copulation attempts (Rosen and Tucker, 1961). Females of several species have a black spot near the genital opening which is most conspicuous when eggs are present in the ovaries. Ingenious experiments with female models that recorded the imprint of gonopodial contacts demonstrated that the black pigmentation both stimulates and directs the orientation of male thrusting (Peden, 1973).

A detailed laboratory and field study showed that the Gila topminnow, *Poeciliopsis occidentalis*, has two different breeding strategies, used by the two main size-classes of males (Constantz, 1975). Larger males establish territories and vigorously court females with conspicuous head-up displays, nibbling at her head and genital area, gonopodial swinging and thrusting, and quivering. Smaller males that are sexually mature but unable to acquire territories because they are dominated by the larger males, wait for an opportunity to dash in and "sneak" a

**Table 5.2.** Qualitative comparison of the mating strategies of territorial and sneak male *Poeciliopsis occidentalis*. (From Constantz, 1975)

| Category | Territorial ♂ ♂ | Sneak ♂ ♂ |
|---|---|---|
| a) Body size | Large | Small |
| b) Agonistic competition | Aggressive | Submissive |
| c) Body colour | Jet black | Light coloured |
| d) Tactile information | Appeared to gain information about ♀ via head and anal nibbling | Appeared to gain no tactile information before gonopodial thrusting; seldom observed anal nibbling |
| e) Courtship | Head-up display | Never observed courting ♀♀ |
| f) Copulatory attempts | Often prolonged bouts of gonopodial thrusts | Fleeting, usually a single gonopodial thrust |
| g) Gonopodial length | Relatively short gonopodium | Relatively long gonopodium |

quick copulation attempt. Table 5.2 summarizes the differences between small and large males. Note that the former have relatively longer gonopodia.

An important competitive advantage for the larger, territorial males is the differential response by adult females to approaches by the two types of males. Females responded to the elaborate courtship of large males by holding position and frequently allowing them to make direct tactile contact. They often fled from an approaching small male. Also, the small males approached a female in a hesitant manner, followed either by fleeing, if attacked by a nearby territorial male, or suddenly dashing to the female and attempting copulation (Constantz, 1975). The latter type of mating attempt may well be less effective in transferring sperm to the female's genital tract than the more deliberate thrusts by larger males.

*Comments.* The poeciliids are small fishes in which the heavy mortality of eggs and early juvenile stages common to most fishes is reduced by the reproductive strategy of ovoviviparity. When the offspring are released from the female they are already active, mobile fish, capable of using a variety of anti-predator mechanisms, including schooling and seeking shelter in vegetation. Average brood size is small, because an adult female can only carry a small number of progeny at once. The poeciliids (along with other ovoviviparous fishes) represent one extreme along a continuum of reproductive strategies in which the number of young produced tends to be inversely related to the duration and effectiveness of parental care.

Chapter 6

# Parental Behaviour

Parental behaviour is any activity performed by one or both parents, after spawning has occurred, that contributes to the survival of their offspring. This does not include cases such as the viviparous fishes, where the female carries her developing young internally for some time before releasing them, or salmonids and others that cover their fertilized eggs with gravel and then desert them. Even though such activities do protect the young, I shall limit the term parental behaviour to the active care and protection of the offspring after spawning.

Such behaviour is widely, but not uniformly distributed among fishes (Keenleyside, 1978b). Of the nearly 250 families discussed by Breder and Rosen (1966), approximately 77% show no parental behaviour at all, about 19% include species that care for the eggs only, and less than 5% have species that care for both eggs and newly hatched young. Furthermore, in families showing parental care, over 75% have active care of the young restricted to the male parent only.

However, these generalities are somewhat misleading because the number of species and the particular strategy used vary widely among the families of fishes with parental care. The best illustration of this is the Cichlidae, a large family in which prolonged parental behaviour is universal. In the New World tropics and subtropics, all species of the genus *Cichlasoma* (over 80 recognized species, Miller, 1966) are biparental (Barlow, 1974a), whereas in Africa the much larger genus *Haplochromis* contains only maternal mouthbrooding species (Greenwood, 1974).

In this chapter I deal separately with parental behaviour that is directed at eggs and at newly hatched young. The former includes only the behaviour of the adult fish, whereas the latter is a more complex, two-way interactive system, in which appropriate stimuli and responses by the parents and their free-swimming offspring are required to maintain the integrity of the family unit.

## 6.1 Care and Protection of Eggs

Two distinct functions are served by the parental care of eggs among fishes: promotion of normal growth and development, and protection against predation. The motor patterns serving these functions depend on whether the eggs are maintained at one location or are carried about by their mobile parents.

### 6.1.1 Eggs at One Location

Species in this category either clear or excavate a spawning site on the substrate, and attach their adhesive eggs to it, or build a nest into which they place the eggs before or immediately after fertilization (see Chap. 4). In either case, proper egg development depends on a variety of parental activities including ventilating, cleaning, removal of dead eggs, and predator repulsion.

## Behaviours Used in Egg Care

*(a) Fanning.* This is the commonest ventilating and cleaning activity. Typically, the fanning parent stations itself close to the egg mass and directs water over it by rhythmic, alternate beating of the pectoral fins. Regular beats of the caudal fin compensate for those of the pectorals, so that the fanning fish holds a constant position despite the vigorous fin movements. The soft-rayed, posterior parts of the dorsal and anal fins also move rhythmically, contributing to stability in position of the fanning fish.

Fishes that lay relatively small clutches often fan their eggs from a position to one side of the clutch. The pectoral fin closest to the eggs moves with wide-amplitude beats; the other moves at the same frequency but with smaller amplitude. The inner fin is clearly more effective in moving water over the eggs (Bergmann, 1968; Baylis, 1974). Fanning towards one side may be accompanied by slight curving of the fish's longitudinal axis, with concave side towards the eggs, as in *Aequidens paraguayensis* (Timms and Keenleyside, 1975), or by the fish rolling slightly to one side with the dorsal surface directed towards the eggs, as in *Chromis dispilus* (Russell, 1971). These postural adjustments are no doubt a result of the fish maintaining a constant position relative to the clutch while fanning vigorously with only one pectoral fin.

Those fish that ventilate their eggs inside enclosures such as caves, burrows, or empty shells, direct water over the eggs in several ways. For example, the toadfish (*Opsanus beta*) spawns in empty conch shells and the male lies on the bottom near the eggs, fanning water over them with the large pectoral fin on that side (Breder, 1941a). The shanny (*Blennius pholis*) spawns in an excavation under a stone, through which the male creates a vigorous current by prolonged tail beating (Qasim, 1956). The male *Cottus gobio* uses his tail and pectoral fins to create currents over the mass of eggs inside his shelter, while lying on the bottom beside or under the eggs. He may even station himself at the shelter entrance, ready to attack intruders, while fanning his clutch with the inner pectoral fin (Fig. 6.1).

The fanning motor patterns may vary among closely related species. For example, the male threespine stickleback (*Gasterosteus aculeatus*) fans his eggs from a position just outside the nest entrance while oriented slightly head-down. Potassium permanganate crystals placed near such a fish clearly show that his fanning movements direct water into and through the nest (Fig. 6.2). Similar orientation and movements have been described for fanning male *G. wheatlandi* (McInerney, 1969), *Pungitius pungitius* (Morris, 1958), and *Culaea inconstans* (McKenzie, 1974). In contrast, the parental male fourspine stickleback, *Apeltes quadracus*, makes two holes opposite each other at each level of his multi-tier nest, and moves a current of water over the eggs by inserting his snout into one of these holes and sucking water towards him by rapid pumping movements of the opercula and branchiostegal membrane (Rowland, 1974a).

*(b) Splashing.* A highly specialized type of egg care occurs in the South American spraying characid, *Copeina arnoldi* (now *Copella* sp., Roberts, 1972). This species

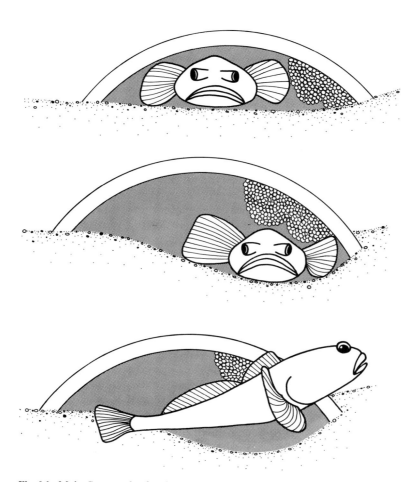

**Fig. 6.1.** Male *Cottus gobio* fanning eggs from different positions at his nest. (After Morris, 1954)

spawns above water on the underside of leaves of emergent aquatic plants. The male periodically swims to the water surface below the egg mass, bends his body into a sigmoid posture, then rapidly turns and flicks water at the eggs with his tail, thus keeping the eggs moist (Fig. 6.3). The cues that males use to direct their splashes are of some interest, given the unusual location of the spawn. By shifting, rotating, and removing parts of a green, circular, artificial spawning disc, Krekorian and Dunham (1972b) determined that splashing direction was controlled primarily by cues from the edge of the spawning surface nearest the eggs. This may be a specific adaptation to the maintenance of eggs on emergent vegetation, since leaf edge cues are probably easier to see than cues from transparent eggs (Krekorian and Dunham, 1972b). Observations of free-living *Copella* in Guyana demonstrated the influences of local weather conditions, especially humidity, on egg-splashing behaviour (Krekorian, 1976). Parental males did not splash during rain and splashed at reduced frequency for several hours after rain stopped.

126

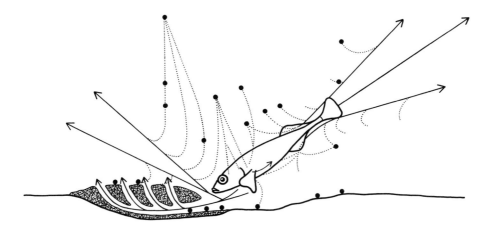

**Fig. 6.2.** Male *Gasterosteus aculeatus* fanning eggs in hist nest. *Solid circles,* positions of KMNO$_4$ crystals; *solid* and *dotted lines,* observed water currents. (After Tinbergen, 1951)

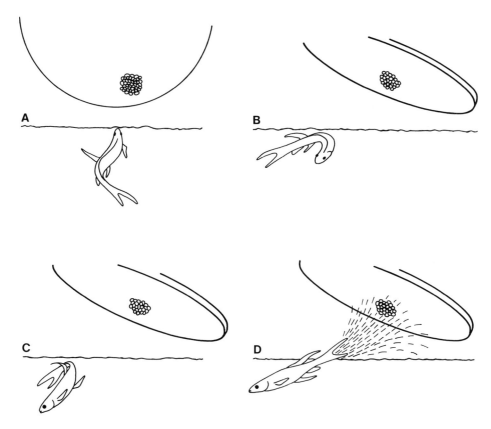

**Fig. 6.3A-D.** Sequence of events during an egg-splashing bout by male *Copella* sp. (After Krekorian and Dunham, 1972a)

This remarkable behaviour appears to be an essential correlate of spawning out of water. It is an example of the extreme specialization in reproductive behaviour to be found in a species-rich, tropical freshwater ecosystem, where selection can be expected to favour new "inventions", such as the use of an aerial spawning substrate, provided the adult fish are able to keep their eggs moist.

*(c) Egg-Contact Behaviours.* A number of other activities, mostly involving direct contact with the eggs, are used by fishes to ventilate and clean the clutch. For example, the parental male *Badis badis*, while in his spawning burrow, performs *shuddering*, in which vigorous, rapid body and tail undulations create sudden water movements that remove loose particles from the egg mass (Barlow, 1964). The male pomacentrid *Abudefduf zonatus* frequently swims through his clam shell nest, beating his tail against the eggs on the inner surface of the shell (Keenleyside, 1972a).

Most egg-guarding fishes mouth the clutch occasionally. Usually contact is gentle and is repeated several times with the same group of eggs. Its function seems to be to remove foreign particles that have not been dislodged by fanning. Parents also use more vigorous mouth-picking to remove dead eggs or small organisms from the spawn. The potential vigour of this behaviour was shown in a study of the pomacentrid *Chromis dispilus*. When small foreign objects were placed among the eggs the guarding male either carried or pushed them away using his mouth (Russell, 1971). More specialized egg-mouthing is shown by the male sponge blenny, *Paraclinus marmoratus*. The eggs are attached in a mass to the roof of a small cave or burrow, and the male frequently pokes them with his snout, occasionally taking a few eggs in his mouth, pulling them outward from the others and releasing them. The mass of eggs is held together by adhesive threads and the vigorous pushing and pulling apparently serves to aerate the agglutinated mass of eggs (Breder, 1941b).

Some species combine vigorous body, fin and mouth contact behaviours during the care of eggs. The South American catfish *Loricaria parva* male rests on his eggs "...like a hen on her nest, cleaning and washing them with his large sucker mouth and keeping a constant water circulation around them by pumping his body up and down, and by fanning with his pectoral fins" (Friswold, 1944, p. 69). Either parent of the freshwater catfish, *Ictalurus nebulosus*, may settle down on the eggs, spread its pelvic fins, and beat them up and down alternately against the eggs (Breder, 1935).

All of these cases of direct egg contact by parental fish appear to have the same functions as fanning, that is, ventilation and cleaning. Apparently, for many fishes that care for adhesive eggs on the substrate, especially where silt and detritus are common hazards, pulling, sucking and pushing at the egg mass is a more effective way of keeping it clean than is fanning alone.

Finally, there are the highly specialized anabantoid fishes, including species in which the male tends his eggs within an air bubble-nest floating at the surface. In addition to typical pectoral fin fanning, oxygen is supplied to the developing eggs by several other activities, including "opercular aeration", in which the male inspires air at the water surface, swims below the nest and releases a stream of small air bubbles into it through his opercular openings (Forselius, 1957). These join the

128

floating nest and make contact with the eggs. The male also may pick up some eggs and air bubbles in his mouth, make chewing movements, and then release them. Males of some species perform "spouting" or "jetting" behaviour by taking in air, moving below the nest, and then with strong opercular movements squirting bubbles up through the nest so forcefully that a spout or jet of air and water is sent above the water surface (Hall and Miller, 1968; Kramer, 1973). All of these bubble-related motor patterns supply the developing eggs with oxygen, a parental activity with clear survival value in the warm, shallow, often stagnant waters of their natural habitat in southeast Asia (Forselius, 1957).

*(d) Moving the Clutch.* An unusual egg-care strategy is followed by some substrate-brooding New World cichlids. A breeding pair attach their adhesive eggs to a small, movable site, most commonly a single leaf on the substrate litter. The parents fan and mouth their eggs in the typical cichlid manner, but they also frequently move the clutch, by grasping the edge of the leaf in the mouth and swimming backwards with it. This was first described for *Aequidens coeruleopunctatus* in Panama (Barlow, 1974a), and has now been seen in several species of this genus, such as *A. pulcher* in Trinidad (M. Itzkowitz and B. Seghers, personal communication), *A. vittatus* in Surinam (Keenleyside, unpublished observations) and *A. paraguayensis* (Timms and Keenleyside, 1975). The adaptive significance of clutch-moving seems to be twofold. First, it is an anti-predator tactic; egg-guarding *A. paraguayensis* pairs quickly move the leaf with their clutch away from disturbance, such as a predator model (Keenleyside and Prince, 1976). Second, since rapid changes in water level are typical of the spawning habitat of these species (Pearson, 1937; Lowe-McConnell, 1964), it may also be a means of keeping the clutch in water of appropriate depth and speed. A more efficient way to move a clutch of eggs is to carry them, either on the body or in the mouth. Egg-carrying behaviours are considered in the next section.

## 6.1.2 Eggs Carried by One or Both Parents

### 6.1.2.1 Methods of Carrying Eggs

Many species of fish carry their newly fertilized eggs for some time after spawning. This may be a brief period, lasting a few hours at most. For example, the spawning female Japanese medaka, *Oryzias latipes*, extrudes a few relatively large eggs that are immediately fertilized by the male, but remain attached to her genital region by adhesive threads. After a few hours she begins rubbing her ventral surface vigorously against plants, filamentous algae, or the substrate, until the eggs become detached and remain in their new location until hatching (Ono and Uematsu, 1957). In many more species the eggs are carried until they hatch; then the young fish leave the parent, or they are carried for some time longer. A variety of specialized techniques is used to carry eggs (see Balon, 1975, for a more detailed review). These include:

(a) A cluster of eggs is attached to a hook-like projection of the supraoccipital crest in the male of the Indo-Australian genus *Kurtus* (Munro, 1967).

(b) Eggs are attached to the female's ventral body surface in the South American banjo catfishes (Bunocephalidae), where they are first embedded in the soft tissue, and later become individually attached to the body inside a stalked capsule (Marshall, 1965).

(c) Eggs are carried on the ventral surface of the male in the syngnathids (pipefish and seahorses). Within the family there is a gradation from loose, ventral attachment to complete protection within a sealed brood pouch (Herald, 1959; Breder and Rosen, 1966).

(d) Eggs are carried inside the parent's buccal cavity. This is "mouthbrooding", a system used by a variety of fishes, both freshwater and marine (Oppenheimer, 1970).

Whichever specific technique is used, the mobility and security afforded the eggs would appear to make egg-carrying a highly efficient parental strategy for avoiding predation and unsuitable water conditions.

### 6.1.2.2 Quantitative Aspects of Oral Egg-Brooding

Fish that carry their eggs in the buccal cavity make periodic, conspicuous mouth movements, variously labelled: chewing, mumbling, gargling, rolling, and churning (Baerends and Baerends-van Roon, 1950; Morris, 1954; Shaw and Aronson, 1954; Reid and Atz, 1958; Oppenheimer, 1970). These movements agitate the eggs, and it is generally assumed that they keep the eggs free of foreign particles, and ventilate them; that is, churning is analogous to fanning by fishes that maintain their eggs on the substrate (Oppenheimer and Barlow, 1968). Shaw and Aronson (1954) showed that vigorous rotation of the eggs, and not simply the increased movement of water around them caused by churning, promoted egg survival. Thus, churning may prevent disruption of normal development that could be caused by heavy yolk lipids settling in one region of eggs that are not frequently rotated (Fishelson, 1966). Other possible functions of churning are listed in the review by Oppenheimer (1970).

In spite of the widespread occurrence of mouthbrooding, there have been few quantitative studies of the phenomenon. The most detailed concerns the African mouthbrooding cichlid *Sarotherodon melanotheron* (formerly *Tilapia macrocephala*), in which the eggs are picked up by the male immediately after spawning (Oppenheimer and Barlow, 1968). The amount of time spent churning by egg-carrying males decreased for several days after spawning and then rose to a peak on the day of hatching (Fig. 6.4). The males continued to carry their young for several more days, during which time churning decreased. These changes are similar to those in egg-fanning by *Gasterosteus aculeatus* (van Iersel, 1953), and support the conclusion that churning, like fanning, is a parental activity whose frequency is largely dependent on the respiratory metabolism of the developing eggs. After hatching, the young fish, even though still carried in their father's mouth, are soon independently mobile within the buccal cavity, and are presumably increasingly less dependent on parental churning for their respiratory needs.

In the biparental cichlid *Aequidens paraguayensis* churning rates for both parents decline steadily after hatching (Fig. 6.5), again suggesting that developing

130

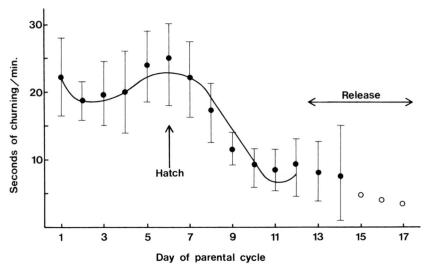

**Fig. 6.4.** Mean daily churning time ($\pm 2$ S.E.) by parental male *Sarotherodon melanotheron*. *Solid circles*, means of 10 males; *open circles*, one male only. (Modified after Oppenheimer and Barlow, 1968)

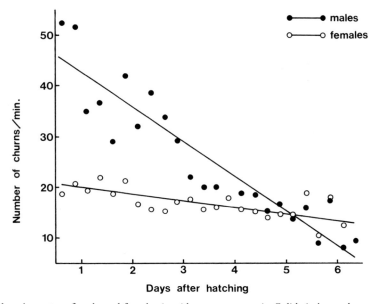

**Fig. 6.5.** Changes in churning rates of male and female *Aequidens paraguayensis*. *Solid circles*, males, *open circles*, females. (Modified after Timms and Keenleyside, 1975)

young are increasingly mobile and less dependent on parental ventilation activities, even while inside their parents' mouths. The reason for the higher male churning rate in this species during the early post-hatch stages is not clear, but the different slopes of the male and female regression lines for churning result in similar rates for both parents by the end of oral brooding (Timms and Keenleyside, 1975).

131

## 6.2 Care of Post-Hatch Young

There is great variety in the extent to which fishes that actively protect their eggs continue to show parental behaviour after the eggs hatch. At one extreme are marine pomacentrids that show none at all. The newly hatched larval fish drift away as part of the plankton, and the parents may soon begin another breeding cycle (Allen, 1975). At the other extreme are the Cichlidae, in which parental care of the brood may extend for several weeks or even months beyond hatching (Burchard, 1967; Sjolander, 1972; McKaye and Barlow, 1976b; Ward and Wyman, 1977). Many fishes are intermediate. The following discussion is organized around two main categories: relatively short-term and relatively long-term post-hatching care of the young.

### 6.2.1 Short-Term Care of Young

The freshwater sunfishes (Centrarchidae) illustrate this condition well. The eggs are guarded by the male parent who continues to drive potential predators away from the nest area after the young have hatched and begun free-swimming. The fry are not very active at first, and tend to swim in schools in the quiet, often weedy waters around the nest site for several hours or days. Strays are snapped up in the parent's mouth and spit into the school. As the young become increasingly mobile, they leave the nest area, feeding and seeking shelter among weeds or rocks (Breder and Rosen, 1966).

A more complex example is the Gasterosteidae (sticklebacks). In European populations of *Pungitius pungitius*, at about the time the eggs hatch, the nest-guarding male parent collects fresh plant materials and accumulates them into an undifferentiated, loosely formed mass, close to the existing nest. This has been called a "nursery" (Morris, 1958), because as the freely swimming young begin to disperse, the parent picks them up and spits them into the new structure, where they remain well hidden. This retrieval and spitting of fry into the "nursery" continues for a few days at most, by which time they scatter and all parental activities cease. "Nursery" construction appears to occur only where *P. pungitius* breeds in dense vegetation. It was not seen in a North American population that breeds along a rocky shoreline with little aquatic vegetation, and where nests are simple collections of algal fragments in excavations under rocks (McKenzie and Keenleyside, 1970).

The parental male brook stickleback (*Culaea inconstans*) converts his nest at about hatching time into a type of "nursery" by "...pulling apart the upper portion of the nest until it is a loose network of material" (McKenzie, 1974, p. 650). Stray fry are quickly retrieved by the male and spit into this loose nest material. Within one or two days retrieval behaviour declines and the fry scatter.

The male *Gasterosteus aculeatus* does not build a "nursery", but as hatching time approaches he pulls apart the upper portion of the nest, thus promoting increased ventilation of the hatching eggs (van Iersel, 1953). As the fry become mobile they school closely above the nest and strays are retrieved by the male for a few days until the parent-offspring bond disintegrates. *G. wheatlandi* males construct a loose "nursery" by adding fresh plant material to the surface of the nest just before the eggs hatch. Retrieval and spitting of fry into the "nursery" continue for a few days (McInerney, 1969).

Finally, available evidence suggests that *Apeltes quadracus* does not change the nest structure, build a "nursery" or retrieve stray young. At least in aquaria the young are ignored by the male parent after they hatch and begin to move from the nest (Rowland, 1974a).

Thus, in some centrarchids and gasterosteids the fry are kept in the nest vicinity for a few days by the guarding male, who actively retrieves strays. In the sticklebacks the elaborate nest that was used to shelter the eggs is no longer suitable for newly hatched fry. Therefore, it is torn apart, and in some species a secondary, looser structure is built. In both families the fry eventually disperse and find shelter near the substrate. The short-term post-hatch parental phase is well-timed to protect the young during the critical period between hatching and independent shelter-seeking.

## 6.2.2 Long-Term Care of Young

In several groups of fishes a close association between parents and their mobile offspring continues for an extended period after the eggs hatch. These include the marine catfishes, Ariidae (Lee, 1937; Gudger, 1916; Atz, 1958), the freshwater catfishes, Ictaluridae (Breder, 1935, 1939), the freshwater bowfin, *Amia calva* (Kelly, 1924; Doan, 1938), and at least one marine pomacentrid, *Acanthochromis polyacanthus* (Robertson, 1973). However, most of the behavioural data on long-term care of the young come from the Cichlidae.

Behaviourists and aquarists have long used (in English) the terms *egg, wriggler* (or *larva*), and *fry* for the three behaviourally distinct stages in the early life of cichlids. Despite Balon's (1975) plea for more consistent and logical labels for the developmental intervals in all fishes, I use the above terms here because of their convenience and familiarity. Wrigglers are newly hatched fish not yet capable of sustained independent swimming. The name comes from the rapid undulating of the body that begins when the tail breaks through the egg capsule as hatching starts, and continues in bouts separated by quiescent rest intervals until the fish becomes free-swimming. It is then called a fry.

The behaviour of cichlid parents towards their wrigglers depends on the overall parental strategy of the species. Substrate-brooders promote the hatching of their eggs by mouthing them more frequently and vigorously as hatching time approaches. Finally they pick the embryos off the spawning substrate and spit them out, usually into a pit previously excavated nearby. Some speicies (e.g., *Herotilapia multispinosa*) may spit their wrigglers into plants above the bottom, especially when breeding in heavy vegetation (Barlow, 1974a). Special glands on the dorsal surface of the head of the newly hatched young secrete a sticky mucous that hardens into tough, flexible threads on contact with water (Fishelson, 1966; Arnold et al., 1968). The threads anchor the wriggler to the gravel, sand or leaf on which it was placed by the parent (Fryer and Iles, 1972). During development wrigglers are often moved to a new pit, or between plants and a pit, especially in response to disturbance by potential predators. Occasionally a parent picks up one or a few wrigglers, rolls them about in its mouth and spits them back with the others. This appears to have a cleaning function.

In contrast, mouthbrooding cichlids carry their wrigglers continually in their mouths, except for brief periods in some biparental species, such as *Aequidens paraguayensis*, when they are released onto the substrate during exchange of young between parents (Timms and Keenleyside, 1975). Associated with oral brooding is a lack of well-developed, mucous-producing, head glands in the juveniles of these species. However, rudimentary head glands are present in the wrigglers of some mouthbrooding cichlids (Arnold et al., 1968), a fact supporting the hypothesis that oral-brooders are derived from the phylogenetically older substrate-brooding cichlids (Peters, 1965; see Sect.6.4.1).

When cichlid young have developed into free-swimming, mobile fry, a two-way communication system between parents and offspring maintains the integrity of the family unit. If undisturbed, the young usually feed in a loose, slowly moving school, with the parents hovering nearby. When danger threatens, the fry quickly gather around the adults, and either stay close to them in a dense school (substrate-brooders) or enter their mouths (mouthbrooders). In either case the rapid clumping of frightened young appears to be in response to signals from their parents. Experimental work on parent–offspring bonds in cichlids has concentrated mainly on analysis of two sets of cues: those sent out by the adults that influence offspring behaviour, and those from the young fish that influence their parents. These will be discussed separately.

### 6.2.2.1 Responses of Cichlid Fry to Parental Cues

Most studies in this area have centred on visually mediated responses, although two types of evidence suggest that other sensory signals should not be ignored. Firstly, many cichlids breed in water of high turbidity, where visual contact between fry and adults may be lost if they are separated by more than a few centimetres (Fryer and Iles, 1972; Baylis, 1974; McKaye and Barlow, 1976b). Second, acoustic signals are used in agonistic encounters among adult cichlids (Myrberg et al., 1965; Schwarz, 1974), and we might therefore expect to find brooding adults in at least some species communicating with their offspring acoustically. Convincing evidence that fry use non-visual cues to stay with their parents is not yet available.

An undisturbed brooding parent typically moves near the substrate in a slow, jerky manner, with frequent stops, starts, and turns. The fry remain in a loose school following the parent's slow progression. The influence of these movements on the young is clearly seen in biparental species, when the two adults exchange positions with the brood. While approaching or leaving the school the adult swims quickly along a straight path and the fry do not follow it. The parent remaining with the school swims along an irregular path as described, and the fry stay near it (Fig. 6.6).

When a family group is disturbed, the fry reactions depend on the degree of excitement of the adults. A highly alarmed parent may dash about, creating turbulence, in response to which the small fish settle quickly to the bottom and remain still until the adult calms down. Mildly disturbed parents of substrate-

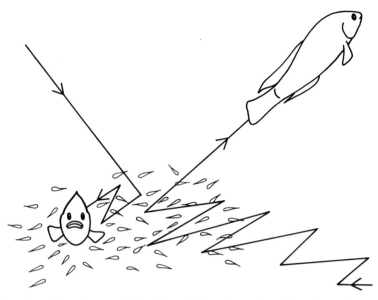

**Fig. 6.6.** Different swimming patterns by parental *Hemichromis bimaculatus* when with and away from their brood. Parent at *left* has just relieved parent on *right*, who swims directly away. (After Baerends and Baerends-van Roon, 1950)

brooding species typically perform "calling" (Baerends and Baerends-van Roon, 1950; Baldaccini, 1973) or "jolting" (Baylis, 1974) movements. In this pattern, the head is jerked slightly sideways, a locomotor wave passes quickly from head to tail, and the pectoral fins make braking movements. The result is a brief jerk or twitch while the fish remains in one spot. "Jolting" is usually repeated several times in rapid succession, and to the observer it is a clear, distinct movement. The fry usually respond by clumping slightly more closely together near the "jolting" adult.

Among mouthbrooding cichlids the parent signals its brood by pitching slightly head-down and swimming slowly backwards. The fry respond by quickly swarming around the parent's head and trying to enter its mouth. They may also contact other parts of the parent's head, such as the opercula, base of the pectoral fins, and the eyes (Fig. 6.7). The backward swimming pattern is made more conspicuous in some species by additional components. For example, *Hemihaplochromis philander* bobs slowly up and down while backing-up (Ribbink, 1971), and *Pelmatochromis guentheri* adults twitch the body sideways and snap the pelvic fins open and shut as they swim backwards (Myrberg, 1965).

The importance of body movement of parental cichlids in eliciting approach by their fry has been amply demonstrated by experiments with models. Working with several species Baerends and Baerends-van Roon (1950) found that shape, size, and surface texture were not important features for stimulating approach, but that movement was. A great range of models elicited approach and following, provided they were moved in the typical slow, jerky manner of an undisturbed parent. *Nannacara anomala* fry responded more vigorously to parental models moved in the jerky manner of a brooding adult than to slower-moving models (Kuenzer,

**Fig. 6.7.** Young of a mouthbrooding cichlid swarming around the parent's head. (After Baerends and Baerends-van Roon, 1950)

1968). *H. philander* fry approached pencils, rubber tubing, small light bulbs and fish models of many colours and shapes, provided they were moved like a parent signalling her brood (Ribbink, 1971). Similar results have been found with several other species (e.g., Kühme, 1962; Kuenzer and Kuenzer, 1962; Brestowsky, 1968; Bauer, 1968).

Some cichlids signal their young by flickering their pelvic fins rapidly open and shut. The visual effect is enhanced by the size and colour of the fins. For example, adult *Aequidens paraguayensis* have large pelvic fins with the most anterior rays extended and conspicuously white. Parents disturbed while their brood is free-swimming repeatedly snap these fins open and shut. The fry usually respond by quickly entering the signalling parent's mouth (Timms and Keenleyside, 1975). A more striking case of pelvic fin-signalling occurs in the substrate-brooding Asian cichlid *Etroplus maculatus*. While guarding free-swimming fry, both parents have a bright yellow body and jet black pelvic fins that they flicker rapidly (up to four flickers per second) when alarmed. When a parental model with movable pelvic fins was placed at each end of a long trough containing *E. maculatus* fry, the fry consistently stayed in the end nearest the model with flickering fins (Cole and Ward, 1970). The degree of clumping of the fry was directly related to the fin-flickering rate (Cole and Ward, 1969).

Many adult cichlids become intensely coloured during the breeding cycle, especially when brooding fry. Thus, colour patterns may be related not only to sexual behaviour between adults, but also to parent–offspring communication. For example, *Herotilapia multispinosa* is usually dull yellowish-gold above and silvery below, with an overlying dark pattern of vertical bars and a mid-lateral stripe. When a breeding pair is guarding fry they take on a striking reverse countershading; the dorsal surface is bright yellow and the sides, ventral surface, and pelvic fins are jet black. These fish often breed in turbid flood plain pools of Nicaragua and Costa Rica, and the conspicuous colour pattern of fry-guarding adults may be an adaptation to allow the fry to maintain visual contact with their parents (Baylis, 1974).

136

Much experimental work has been done with models to examine the responsiveness of cichlid fry to parental colours. In an early study Noble and Curtis (1939) found that *Cichlasoma bimaculatum* fry responded most strongly to black and to blue models, while *Hemichromis bimaculatus* fry responded preferentially to red models. Similar results with the same species were obtained by Baerends and Baerends-van Roon (1950). The biological significance of these findings is that the colours evoking the strongest approach responses are the predominant colours of breeding adults of those species. Clearly the young fish were reacting to the models as though they were their real parents.

Similar results have been recorded for a variety of other cichlids. For example, fry of the South American dwarf cichlids *Apistogramma reitzigi* and *A. borelli* responded most strongly to all-yellow and to black and yellow models respectively; these are the predominant colours of the females of the two species while brooding fry (Kuenzer and Kuenzer, 1962). The females of another dwarf cichlid, *Nannacara anomala*, are black with a complex overlying pattern of white spots and patches. Models with white-on-black patterns produced the strongest following responses by *N. anomala* fry of a wide range of models (Kuenzer, 1966, 1968).

The role of contrast in brightness between a parental model and its background when presented to fry, and between a dark spot and the area surrounding it on a model were examined in *N. anomala* and *Hemihaplochromis multicolor* respectively. In both cases greater contrast elicited stronger responses (Kuenzer, 1968, 1975).

Finally, the influence of adult size on the responsiveness of fry was studied in two mouthbrooding species, *Tilapia nilotica* and *H. multicolor* by Langescheid (1968) using glass ball models. *H. multicolor* breeds at a much smaller size than *T. nilotica*. The range in diameter of models eliciting strong approach and contact behaviour was from 2.5 to 8.0 cm for *T. nilotica* and from 1.0 to 2.5 cm in *H. multicolor*, thus supporting the concept that the fry of these species use relative size of adults as one of the visual cues by which they selectively approach and contact their own parent.

In summary, a variety of visual cues emanating from parental cichlids may be used by free-swimming fry to maintain proximity to their parents. These include orientation, swimming movements, pelvic fin flickering, colour patterns, contrasts in brightness, and size of adults. Any attempt to establish the relative importance of different visual cues for a species should be based on a thorough knowledge of its natural ecological setting. For example, the simple fact of variations in turbidity of natural waters is usually ignored in laboratory experiments, and yet Baylis (1974) has shown that the conspicuous colour patterning of brooding *Herotilapia multispinosa* becomes understandable only when one is familiar with the turbid conditions of the natural breeding sites.

## 6.2.2.2 *Direct Contact of Parents by Fry*

Free-swimming fry of mouthbrooding cichlids contact various parts of the parent's head when trying to re-enter the mouth (Fig. 6.7). In response, the parent opens its mouth and the fry quickly swim inside. However, the fry of some substrate-brooders make a different type of physical contact with their parent's body. This was first reported for the discus fish, *Symphysodon discus*, in which the young actively feed on the copious mucous produced during the brooding period by glands in the parent's skin (Hildemann, 1959). Feeding on parental mucous has since been observed in other cichlids (e.g., *Pterophyllum scalare* (Chien and Salmon, 1972)), and it can be induced in many substrate-brooding species by depriving the fry of other sources of food (Barlow, 1974a).

Detailed investigations of parental contact behaviour by *Etroplus maculatus* fry have been conducted by Ward and his colleagues over several years. In this species two distinct types of contact can be distinguished (Quertermus and Ward, 1969). One has been termed "glancing", in which the young fish approaches the parent's side, makes brief, rapid contact, either with its opercular region or the side of its own body, and then bounces away. The contact is often repeated several times in rapid succession. The second pattern is the "micronip" which consists of the fry directly approaching the side of the parent, making brief oral contact and then either turning and swimming away or bouncing straight backwards. In the latter case, it may immediately move forward and make another mouth contact. Micronipping involves the ingestion of mucous secreted by cells in the parental epidermis (Ward and Barlow, 1967).

Analysis of temporal changes in frequency of both contact behaviours as young *E. maculatus* develop has provided information on their functions. Micronipping is primarily a mucous feeding pattern; its frequency is highest during the first few days of free-swimming (Quertermus and Ward, 1969), when the number of mucous glands in the parental epidermis is at a maximum (Ward and Barlow, 1967). Glancing is relatively rare during the first week of free-swimming, but then rapidly increases in frequency until the fry are about 18 days old. The latter period coincides with their increased mobility, and glancing may help to maintain the integrity of the family unit. The frequent performance of glancing probably keeps most of the fry near their parents, and the tactile contacts by the glancing fry may stimulate parental motivation in the adults (Ward and Barlow, 1967; Ward and Wyman, 1977).

Parent-contacting behaviour is also common in aquarium-held *Cichlasoma citrinellum*, where it is primarily a feeding response; fry which have recently been fed perform consistently fewer parental contacts than unfed fry, and there is a direct relationship between the time since feeding and the frequency of contacts (Noakes and Barlow, 1973a). Increased contact behaviour in *C. citrinellum* is positively correlated with the number of mucous cells per unit area of epidermis, and negatively correlated with epidermal thickness of parent fish; the latter relationship may be caused by the frequent removal of epidermal tissue by contacting fry (Noakes, 1973).

The extent of parent-contacting as a feeding response in nature is unknown, except for *Etroplus maculatus* and *E. suratensis*. These species have been carefully observed in their natural habitat in Sri Lanka, where the young of both species frequently glance and micronip at their parents throughout the period of their close association, which may last for several months (Ward and Wyman, 1977). The fact that fry of many captive Central American cichlids graze on parental mucous only when deprived of alternative food (Barlow, 1974a) suggests it will not be common in their natural habitat, where food for small fish is unlikely to be scarce. On the other hand, *S. discus* fry may ingest mucous from their parents' bodies for several weeks, even when alternative food is abundant (Hildemann, 1959). Clearly this is a widespread pattern among cichlids, but its significance for free-living fish will only be determined by further studies in the field.

The non-feeding contact behaviours, such as glancing, may help to maintain parental motivation in the adults, and group cohesion among the fry, as mentioned above. In addition, Wyman and Ward (1973) argue that glancing and micronipping in *E. maculatus* are the juvenile motor patterns from which many adult agonistic and courtship patterns are derived ontogenetically. They describe the contact behaviours gradually changing into a series of contact and non-contact social behaviour patterns in the developing adults. It remains to be determined if this hypothesis can be generalized to other species in which the juveniles regularly make physical contact with their parents.

Whereas most known cases of parental contact by free-swimming fry involve cichlids, similar behaviour has been observed in at least one non-cichlid. At Heron Island (Great Barrier Reef) the fry of the pomacentrid *Acanthochromis polyacanthus* stay with both parents for several weeks after hatching. During this extended period of family life (unusual for pomacentrids) the fry occasionally make vigorous contact with the bodies of their parents, and also with other small fishes. The behaviour resembles the feeding contacts of cichlids, but no proof of mucous ingestion by the young *A. polyacanthus* was obtained (Robertson, 1973).

### 6.2.2.3 Responses of Cichlid Parents to Progeny Cues

Many cichlid fishes breed at times and places where they are likely to encounter other breeding cichlids. While substrate-brooders are guarding schools of mobile young, the latter may be scattered by predator attacks or by vigorous aggressive encounters between their parents and other brooding adults (McKaye and McKaye, 1977). If fry become separated from their family groups, must they find their own parents again to survive, or will they be accepted into schools of young being guarded by other adults? The strong possibility of this problem arising, especially with species that produce many hundreds of young with each spawning, and that live in water of high turbidity (McKaye and Barlow, 1967b), has led to a number of laboratory studies of the ability of substrate-brooding adults to distinguish between their own and strange young. Results have varied, even with the same species, partly due to variation in techniques of handling the young fish. Nevertheless some generalizations have emerged.

Adults of several species do distinguish between their own fry and those of other species; the latter are usually eaten, either immediately or a few hours after being placed with a foreign brood (Noble and Curtis, 1939; Baerends and Baerends-van Roon, 1950; Myrberg, 1964, 1966). However, adults will often accept conspecific fry from other broods into their own brood. The age (and hence size) difference between the introduced and resident fry is the critical variable. Generally, strange conspecific fry are accepted by brooding adults if they are younger than their own, but not if they are older (Noble and Curtis, 1939; Myrberg, 1964). Adult brooding *Cichlasoma citrinellum* accepted conspecific fry which were 2, 10, 13, and 35 days younger and 2 days older than their own, but not fry which were 10 or 13 days older (Noakes and Barlow, 1973b). In the latter study the readiness to behave parentally towards strange fry younger than their own was dramatically demonstrated by one pair of adults that was induced to continue fry-guarding for nine months by replacing their brood at about monthly intervals with a new batch of one- to two-week old fry.

However, some cichlid species appear to be less discriminating than others in their readiness to accept strangers into their broods of young. For example, *Nannacara anomala* will accept a wide range of progeny substitutes. This is a substrate-brooding species in which the female provides most or all of the brood care. If her own young are replaced by artificial eggs, *Tubifex* worms or *Daphnia* (at the egg, wriggler, and fry stages respectively), or by young *Hemichromis bimaculatus* at the same developmental stage as her own, the female *N. anomala* shows parental behaviour of appropriate form and intensity towards all of these substitutes (Kuenzer and Peters, 1974). Further, she continues to perform appropriate parental behaviour, although often of lower intensity, when her entire brood is replaced by a conspecific brood at a different developmental stage (older or younger).

All of the above results were obtained with aquarium-held fishes. A recent experiment in Lake Jiloa, Nicaragua, has confirmed the age-specificity of acceptance of foreign fry by brooding cichlids. Fry from other conspecific broods were introduced into schools of young *C. citrinellum*, and were accepted by the guarding parents as long as they were about the same age as their own young (McKaye and McKaye, 1977). Fry that were 9 to 16 days older than the host brood were attacked by the parents; those 19 or 20 days younger were eaten by the host fry. That adoption of young from other conspecific broods occurs naturally in Lake Jiloa was shown by the gradual increase in size of some broods of *C. citrinellum, C. nicaraguense, C. longimanus*, and *Neetroplus nematopus*. Some of these increases were due to active "kidnapping" of fry by the parents of another brood.

The above work leads to the obvious question: What are the cues by which adult cichlids distinguish their own from strange young? Present evidence suggests that chemical and visual cues are most important. Kühme (1963) used a simple choice apparatus to show that female *H. bimaculatus* responded preferentially to water coming from a container holding their own wrigglers or fry over that coming from other conspecific broods. Chemicals in the inflowing water must have been the critical cues. A similar distinction was not made between their own and other eggs. Using similar apparatus, Myrberg (1975) found that female *C. nigrofasciatum* distinguished between water from their own eggs and that from eggs of other species, during the 10–15-h period before hatching. This suggested that the pertinent chemical cues are released only as the egg capsules begin to disintegrate.

*C. nigrofasciatum* females were also able to distinguish between their own and heterospecific wrigglers on the basis of chemoreception. Adult males showed a similar, but weaker response to chemicals from their own wrigglers, a sex difference corresponding to the greater parental role of the female in this species (Myrberg, 1966, 1975). When given visual access only to broods of wrigglers, female *C. nigrofasciatum* did not discriminate between their own and those of *H. bimaculatus*, but when visual and chemical cues were both available, the females responded more strongly to their own. Thus it appears that in *C. nigrofasciatum* visual cues serve to orient the parents towards broods of wrigglers, while chemical cues are used for

140

discriminating between their own and foreign wrigglers. When the brood is at the fry stage, vision becomes increasingly important for the parent, not only to maintain contact, but to discriminate between fry of different broods (Myrberg, 1975).

Taken together, the laboratory and field work suggest that visual cues associated with size are most important in enabling brooding adults to distinguish between their own and foreign free-swimming fry. Smaller ones may be treated as food by the host fry, larger ones are treated as potential predators of the host fry and are chased away. Those about the same size may well be accepted, presumably because a few strangers among a brood of several hundred can escape detection.

### 6.2.2.4 Communal Care of Young

The cooperative, joint care of young by individuals in addition to the true parents has seldom been reported for fish, although it is common among some groups of birds and mammals (Brown, 1975; Wilson, 1975). Recently the phenomenon has been observed among cichlid fishes, both in aquaria and in nature. *Lamprologus brichardi* is a substrate-brooding, biparental species endemic to Lake Tanganyika. In aquaria the young of one brood are tolerated inside their parents' territory even after one or more subsequent broods have been produced. They mouth the eggs and move freely among their younger siblings, although the parents aggressively keep other adults away from the young (Coeckelberghs, 1975). Similar cooperative brood care behaviour in this species has been seen in its natural habitat (Brichard, 1975).

As yet no data are available from the field by which to assess the functional significance of this behaviour. Unless the members of different broods can be distinguished from each other, an observer cannot determine the degree of relatedness of the cooperating adults. This information will be necessary before it can be unequivocally stated that older siblings "help" their parents to raise subsequent broods, as is well-documented for some birds and mammals (Brown, 1975).

Another form of communal parental behaviour is the joint care by several brooding adult pairs of their combined schools of offspring. Ward and Wyman (1977) found several large schools of young *Etroplus suratensis* in Sri Lanka being actively guarded by more than two adults. The fry showed glancing and micronipping behaviour against all the brooding adults. Although there was no way of proving that each adult was the parent of some young in the school, the behaviour of adults and fry strongly suggested that separate broods had coalesced into a single large one that was being jointly guarded by all their parents. McKaye and McKaye (1977) observed three adult pairs of *C. citrinellum* communally guarding a large school of fry in Lake Jiloa, Nicaragua. Occasionally the three pairs returned to their separate shelter caves among the rocks, each followed by some of the fry. Repeated separation and union in this manner would undoubtedly mix the broods.

A somewhat different situation was observed by Sjolander (1972) in Nigeria. Mixed broods of juvenile *Tilapia mariae* and *T. zillii* were being guarded by adults of both species. As in the case of *E. suratensis*, it appeared that the mobile schools of young *Tilapia* had combined and that the adults were jointly protecting their mixed-species broods. Similarly, mixed schools of young *T. mariae* and *T. melanopleura* were seen being guarded by adults of either species in the Benin River, Nigeria (Burchard, 1967).

One explanation for these cases of communal brooding may be that schools of young conspecifics of about the same age are strongly attracted to each other. Certainly this is the case with many cichlids held in captivity. If two adult pairs of one species have broods of similar age in a single aquarium, the mobile fry will try to combine whenever they approach each other. If the fry are all about the same size, their appearance may be so similar that it is difficult for the adults to distinguish their own offspring, and thus it may be easier for them to collaborate with each other in guarding a combined school than to separate their own young from the others. A similar explanation may hold for the communal guarding of mixed-species broods seen in nature.

An interesting case of interspecific "foster-brooding" has been described among the maternal mouthbrooding cichlids of Lake Malawi (Ribbink, 1977). Females of three rock-inhabiting, predaceous species (*Haplochromis macrostoma*, *H. polystigma* and *Serranochromis robustus*) were seen guarding mixed broods consisting of their own fry and some fry of a near-surface plankton-feeding cichlid (*H. chrysonotus*). The latter were positively identified by capturing a mixed brood and rearing it in captivity. No information is available on the transfer of young from their mother to the "foster mother", but the phenomenon was not an isolated case. Over a 10-day study period, all individuals of the three predaceous species with young were seen to have mixed broods (Ribbink, 1977). The explanation may be that whenever the *H. chrysonotus* fry become separated from their mothers and reach the substrate they are attracted to the foreign broods, where they benefit from protection by the "foster-mothers". Parental motivation of the latter may be so high that they are inhibited from attacking strange fry of about the size of their own. Further detailed observations in the field are needed to test these suggestions.

## 6.3 Male–Female Roles in Parental Behaviour

Among biparental fishes, comparison of the contributions to care of the young by male and female parents has been restricted mainly to cichlids. Most biparental cichlids are substrate-brooders, and in general the differentiation of male–female roles is most pronounced when the progeny are at the egg stage. Typically, the female is said to perform more of the direct egg-care behaviours and the male is the more active defender of the brood against potential predators (Baerends and Baerends-van Roon, 1950; Barlow, 1974a). However, this generalization is based mainly on those species in which male–female parental roles are strongly differentiated, and quantitative data are not needed to confirm the division of labour.

142

In many biparental cichlids, brood-care behaviours are more equally shared between the sexes. For example, female *Pterophyllum scalare* parents spent more time fanning their eggs than males did on the first day after spawning, but there were no sex differences on days two and three (Chien and Salmon, 1972). In both *T. mariae* (Baldaccini, 1973) and *H. multispinosa* (Smith-Grayton and Keenleyside, 1978) females performed more egg-fanning bouts than males did, but in the latter species the total duration of fanning per unit time did not differ between the sexes. The *H. multispinosa* parents did not simply exchange duties over the eggs; the female frequently performed more than one distinct fanning bout before her mate relieved her. Mouthing of the eggs also did not differ between the parents in *P. scalare* and *H. multispinosa*. On the other hand, *Aequidens paraguayensis* females fanned more and mouthed the eggs much more than did their male partners (Timms and Keenleyside, 1975).

After the eggs have hatched, the separation of parental roles is also not uniform across species. Wrigglers are often transferred orally between substrate pits (*H. multispinosa*) or leaves (*P. scalare*), and individuals that become separated are picked up and spit back into the mass; both parents share more or less equally in these tasks. *H. multispinosa* adults also frequently pick up a few wrigglers, roll them around in the mouth and gently spit them back with the others. This was done more often by female than male parents (Smith-Grayton and Keenleyside, 1978).

Once their progeny reached the free-swimming fry stage male *T. mariae* did more pelvic fin-flickering to attract the young than females did (Baldaccini, 1973). Cole and Ward (1969) found that male and female *Etroplus maculatus* did not differ in frequency or duration of pelvic fin-flickering, as measured on days 1, 3, 6, 9, and 12 post-hatching. Also, both fry-guarding *P. scalare* parents flickered their enormous dorsal and anal fins with about equal frequency (Chien and Salmon, 1972). Retrieval of fry that have strayed from the school was shared equally between parents in *P. scalare* and *H. multispinosa*.

The generalization that the male is more aggressive than the female when a pair is guarding their young is also not consistently supported by the data. In the highly aggressive *C. nigrofasciatum* both members of brooding pairs are about equally aggressive towards potential predators (Krischik and Weber, 1975; FitzGerald and Keenleyside, 1978). Male *H. multispinosa* chased intruders away from the brood more frequently than females did, but only when their young were at the egg stage (Smith-Grayton and Keenleyside, 1978). In one of the few quantitative field studies on cichlid parental behaviour Cichocki (1977) found that the female of a brooding pair of *Biotodoma cupido* performed many more aggressive acts than the male did, while they were both guarding eggs and wrigglers.

Thus it seems clear that among biparental cichlids both parents are capable of all brood-care behaviours, and that separation of roles between them occurs in some species, and at certain stages of their progeny's development. However, the experimental data by no means support a simple division into brood care by the female and aggression by the male. Data collected from free-living fish will be useful in interpreting the quantitative results of aquarium work. The value of field observations for such comparisons has been discussed recently by Fernald and Hirata (1977).

## 6.4 Evolution of Parental Behaviour

Three major types of variation can be seen in the parental behaviour of fishes. These are variation in: (1) duration of parental behaviour during the development of a single brood; (2) degree of protection afforded the young by their parents; (3) parental roles of males and females. The first two of these appear to follow progressive trends, representing increasing lengths of time during which the young receive parental care, and increasingly effective means of providing care respectively. It is more difficult to see a clear progression in male–female parental role sharing.

Since intensive and prolonged guarding of the young undoubtedly contributes to their survival, one might ask why selection has not favoured its development more widely among fishes. Presumably the answer lies in the complex of physical and biotic factors that determine the success of any population. Relations between adults and their offspring are only one of these determining factors. Species with no direct parental care have other reproductive strategies for ensuring success. These include rapid insertion of fertilized eggs into the plankton (many marine species), production of very large numbers of eggs in a restricted zone (Pacific herrings), and desertion of buried eggs (Pacific salmon). Thus, parental behaviour should be seen as an adaptation permitting many species to occupy habitats that might otherwise be highly unfavourable to survival of their young.

Fishes that practise parental care are not randomly distributed in aquatic ecosystems. Typically they breed on or near the bottom in shallow water. Most species modify a small area of the substrate as a site in which to spawn and rear their young. Protecting the brood against predation would seem to be the primary function of parental care, although providing the developing eggs with adequate oxygen is also important, hence the almost universal presence of egg-aerating techniques among parental fishes. In fact, the major exception to the close association of nesting sites with the substrate is shown by fishes breeding in freshwater pools and swamps where oxygen levels may become very low. Many of these species build floating nests of air bubbles to hold the eggs that then have better access to surface oxygen than if they were maintained in substrate nests (Roberts, 1972). The best-known examples are the anabantoids, but various other genera also build floating nests (e.g., *Gymnarchus, Callichthys,* and *Hoplosternum*). All of these breed in quiet, shallow, freshwater swamps and pools (Breder and Rosen, 1966).

### 6.4.1 Evolution of Oral Brooding of Young

Oral brooding by the parents would seem to be the most efficient method of protecting offspring. To get at the small fish, highly specialized predatory techniques may be required. For example, within the large African genus *Haplochromis*, all members of which are maternal mouthbrooders, at least eight species feed primarily on the eggs and wrigglers of other *Haplochromis* (Greenwood, 1974). The method by which these prey are obtained has not been well established; the predators may suck them out of the parent's mouth (Lowe-

McConnell, 1969) or the brooding females may "voluntarily" jettison their broods when harassed by persistent predators (Fryer and Iles, 1972).

Given the apparent advantages of oral brooding, there has been considerable interest in the evolutionary origins of this form of parental care in fishes. Various proposals have been discussed by Oppenheimer (1970). These include: shortage of suitable substrate for the developing eggs and wrigglers, and advantages of a colonial breeding system, where clusters of males maintain territories and spawn repeatedly with different mouthbrooding females. The latter system must be compared with substrate-brooding, where the progeny are more open to predation, but the mean brood size is usually much larger (Fryer and Iles, 1972). Comparison of oral- with substrate-brooding, as a means of understanding the evolution of the former, can best be done with cichlids, in which both parental strategies are widespread. Several types of evidence suggest that substrate-brooding is less specialized, hence more primitive, than mouthbrooding.

In Africa the majority of cichlid species are maternal mouthbrooders, and they mostly have limited ranges. Many are endemic to a single lake, where they have recently evolved; in fact, African lake cichlids form one of the best-known examples of recent, explosive speciation among freshwater fishes (Fryer and Iles, 1972; Greenwood, 1974). Most of the substrate-brooding species are widely distributed across western and northern Africa, suggesting they have a long history on the continent (Lowe-McConnell, 1959). Similar information is not available for New World cichlids. South America is the ancestral origin of all species, including those found in Central America and Mexico. The latter areas are dominated by the large and rapidly speciating genus *Cichlasoma*, together with a few derived genera (Miller, 1966; Myers, 1966). All of the Central American and most of the South American cichlids appear to be substrate-brooders; only a few species in the genera *Geophagus* and *Aequidens* are known to be mouthbrooders, and most of these brood their eggs on the substrate and then pick up the wrigglers for oral incubation (Reid and Atz, 1958; Timms and Keenleyside, 1975). A single species (*Geophagus hondae*) is known to be a true maternal mouthbrooder (Sprenger, 1971; personal observations). The distributions of South American cichlids are not well known in detail, but many species appear to have broad ranges (Lowe-McConnell, 1969; Gosse, 1975). There is no clear evidence that mouthbrooders tend to have one type of distribution and substrate-brooders another.

Morphological evidence on the relative antiquity of the two brooding strategies concerns the adhesive structures on the eggs and newly-hatched young. The eggs of substrate-brooders have adhesive threads on the surface that anchor them to the substrate. Mouthbrooders' eggs have no such threads, or else they are sparse, as in *Sarotherodon galilaeus* and *S. melanotheron* (Trewavas, 1973). The wrigglers of substrate-brooders have head glands that secrete mucous threads, but similar glands are present only in rudimentary form in juvenile mouthbrooders (Peters, 1965). These structural differences support the suggestion that mouthbrooding cichlids are evolutionarily derived from substrate-brooders (Fryer and Iles, 1972; Trewavas, 1973).

Behavioural evidence also supports this trend. Most, if not all, substrate-brooders hasten the hatching process by frequent mouthing of the eggs as hatching time approaches. The wrigglers are picked off the spawning substrate by one or both parents and spit out into a prepared pit or onto nearby vegetation. The parents usually transfer wrigglers from one site to another during their development; such transfers are always made by mouth. As the young become mobile they swim in a compact school that gradually becomes more diffuse as the fry develop. The adults are quick to retrieve stray fry by snapping them up in the mouth and spitting them into the school. Throughout these transfer, an adult that is disturbed while carrying young fish in its mouth may continue to hold them for up to several minutes (Myers, 1939; Lowe-McConnell, 1959; Collins, 1972). It seems likely that under certain adverse conditions (e.g., crowded brooding areas, high predation pressure), selection would favour the extension of brief episodes of oral transport into prolonged carrying of the young, and eventually into complete mouthbrooding, beginning at spawning.

The intermediate brooders (e.g., *Geophagus jurupari* and *Aequidens paraguayensis*) that guard their eggs on the substrate, but orally carry the young after hatching, appear to be in a transition state between the two basic brooding strategies. Behavioural and ecological information on such species from their natural habitats would be extremely valuable in the attempt to understand the evolution of oral-brooding.

The physical characteristics of cichlid spawning and brooding habitats may also provide clues to the origins of mouthbrooding (Roberts, 1972). Typically the oral-brooding species are found in lakes and are much less common in rivers than are the substrate-brooders. In the relatively still and clear inshore waters of some of the African lakes, exposed eggs and young fish near the substrate would be vulnerable to visually oriented predators. On the other hand, many of the larger rivers of Africa and South America are less clear, due to heavy silt loading or to the particular characteristics of the river basin producing "black" or "white" water; in such cases the eggs and young of cichlids would be difficult for predators to detect visually. Thus one would expect to find mouthbrooding a common strategy in the clear lakes, such as Tanganyika and Malawi. In both these lakes most of the cichlids are mouthbrooders, and the exceptions brood their young in well-hidden shelters. For example, *Julidochromis ornatus* and some close relatives in Lake Tanganyika spawn inside small rock caves, where the young are closely guarded for several weeks (Wickler, 1965b; Brichard, 1975). In general, Fryer (1969, 1977) has argued that incorporation of maternal mouthbrooding into the reproductive systems of African lake cichlids has been one of the most important adaptations contributing to the phenomenal success of these fishes.

## 6.4.2 Evolution of Male–Female Parental Roles

Active parental care of the young occurs in about 23% of recognized families of fishes, and in about 75% of these, brood care is by the male parent alone (Breder and Rosen, 1966). The Cichlidae is one of the unusual families, in that both maternal and biparental care are much more common than strictly paternal care. The division according to sex of the brooding parent parallels the division into care by one or both parents. All substrate-brooding cichlid species appear to be biparental; all but a few of the strictly mouthbrooding species have maternal care

146

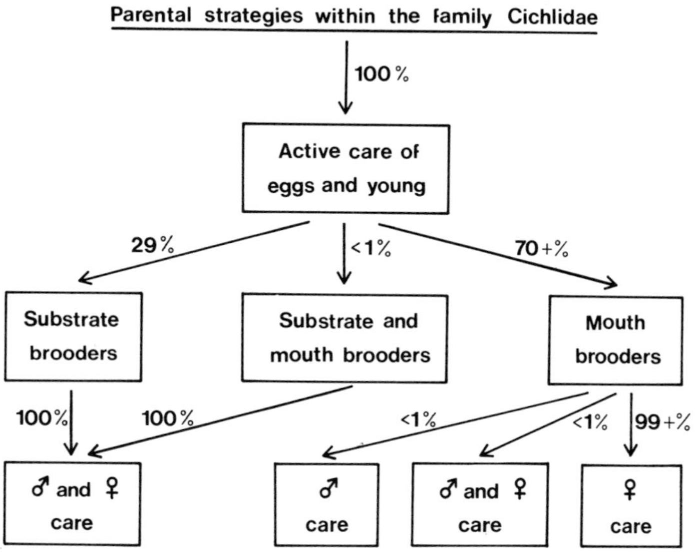

**Fig. 6.8.** Percentage of cichlid species showing various parental strategies

only (Fig. 6.8). Thus the relatively recent trend towards oral brooding of the young has been accompanied by emphasis on brooding by the female parent alone.

Perhaps the major explanation for the association of mouthbrooding with single parent care is that both parents are needed to protect a large brood of eggs, wrigglers, and especially mobile fry that are not held in the mouth. This has seemed obvious to observers watching cichlids in the field (Barlow, 1974a; McKaye, 1977; B. Seghers, personal communication; Keenleyside, observations in Surinam). Cichlids generally breed in areas rich in potential predators, including conspecifics. The extent of this pressure has been documented most dramatically by McKaye (1977) who found that among several biparental cichlid species in Lake Jiloa, Nicaragua, less than 10% of breeding pairs were able to raise their offspring to the stage at which they dispersed from the breeding territory. One must assume that under these conditions a single parent could never raise a brood successfully.

The handicap suffered by a single parent of a biparental species is supported by some experimental evidence. Removal of either member of a brooding pair of *Herotilapia multispinosa* in the presence of potential predators resulted in increased loss of young compared to control situations where both parents remained with their brood (Keenleyside, 1978). Thus, the suggestion of Emlen (1973) and of Williams (1975) that guarding of the young in biparental fishes can be done as effectively by one parent as two, is not supported by the evidence.

Finally, given that single parent mouthbrooding is derived from biparental substrate-brooding, why has this resulted in maternal care so much more often than paternal care among cichlids? In non-cichlid fishes where one parent shows extended care of the young, it is usually the male that takes this role. This is associated with high levels of aggression by the male before and after spawning. He establishes a territory, prepares a spawning-site or nest in it, courts ripe females, and after spawning with one or more of them, actively defends his brood. Sticklebacks and sunfish exemplify this system. Among mouthbrooding cichlids much the same sequence is followed. The male actively defends a territory, in which a breeding site is prepared. He courts females and spawns with one or more in succession. The major difference is that in the cichlids the female picks up and carries away her fertilized eggs; in the other species the female leaves her eggs with the male. In both systems the females are generally smaller and more drably coloured than the males. Selection has presumably favoured development of secondary sexual characters in the males that are used to attract and stimulate females.

However, the question remains. Why has a system arisen among uniparental cichlids where females keep their eggs with them after spawning, while in uniparental sticklebacks and sunfish (and several other groups) females leave their eggs with males to brood? Perhaps the critical distinction is that cichlid breeding habitats exist within highly complex, tropical freshwater ecosystems, where successful defense of egg clutches on the substrate by a single parent is improbable. The sunfish and sticklebacks breed in less complex, temperate ecosytems where single parent defense is possible. In the latter case it is the male, who is already highly aggressive in defense of his spawning site before breeding, who stays with the young and protects them. Female mouthbrooding cichlids move away from the spawning sites with their eggs, and often go to special nursery areas where the young fish are eventually released (Fryer and Iles, 1972). Thus, the cichlid system, especially in the many species where the males' territories are grouped in spawning arenas, resembles the arena system of lek-breeding birds and mammals.

Chapter 7

# Social Organization

The social organization of a population of animals is the sum of the inter-individual relationships among the members of that population (Brown, 1975). An ethologist can only determine the character of these relationships by carefully observing the interactions among individual animals. All such interactions are examples of social behaviour. For the sake of convenience in discussing this complex subject I distinguish between two general aspects of social behaviour in fishes: the *social units* in which fishes occur, and the *behavioural mechanisms* underlying the relationships within and between social units.

*Social Units.* I consider four main types: single fishes, long-term male–female pairs, small groups (composed of adults of both sexes and sub-adults that are usually not offspring of the group adults), and schools.

*Behavioural Mechanisms.* The major ones are:

*(a) Individual Distance.* This is the minimum distance that an animal maintains between itself and other conspecifics (Wilson, 1975). An individual prevents encroachment on this distance either by withdrawing from an approaching animal, or causing it to withdraw. The phenomenon was first closely examined with birds and large mammals in zoos, where it was recognized as an important component of the environment needed to minimize damaging aggression among captive animals (Hediger, 1955). In fishes it determines the spacing among individuals in schools, groups, and pairs.

*(b) Home Range.* This is the area through which an animal or group regularly travels in pursuing its daily activities (Burt, 1943). Typically one part of the home range is used more intensively than the rest. This may include a nest, burrow, or special hiding or feeding place, and is called the *core area.* Although the resident(s) may aggressively exclude intruders from the core area, the home range as a whole is not defended. Neighbours tend to avoid each other rather than fight at areas of home range overlap. The home range is a well recognized phenomenon among mammals and birds (Brown, 1975), but has not been carefully documented for many fishes. Using appropriate techniques for tracking known individuals, data on fish home ranges are now accumulating (Reese, 1975; Stasko, 1975).

*(c) Territory.* This is an area occupied more or less exclusively by one or more animals, by active repulsion of potential intruders through defense or advertisement (Wilson, 1975). The pioneering definition by Noble (1939), that a territory is "any defended area", is still widely used by behaviourists, but the above

expansion by Wilson is an improvement because it combines the two concepts of aggression and exclusive use. Discussions of the problem of defining territory can be found in Wilson (1971, 1975) and Brown (1975). Territoriality is widespread among fishes, predominantly, but not only, during reproduction.

*(d) Dominance.* Any behaviour by which one animal gains priority over others in attaining a shared resource, such as food, shelter or mates, is dominance behaviour. It is common among well established groups, where individuals recognize one another, and often results in a hierarchical system in which certain individuals are consistently dominant over others. Dominance behaviour is common among fishes held in captivity, but clear evidence for it among free-living fishes is limited to a few species.

In what follows I discuss the four basic social units in fishes, together with evidence for the roles played by the four behavioural mechanisms in shaping the characteristics of the units. Examples are mostly from field observations of free-living fishes, because the objective is to look for important ecological correlates of social organization. This should bring us closer to an understanding of the evolutionary origins of the social structures found among living fishes. One complicating factor is that the individuals of a local population may shift from one form of social organization to another during day–night, tidal or breeding cycles (e.g., Hobson, 1968b, 1972; Ehrlich et al., 1977). Shifts also occur with age, as for example in some salmonids that are substrate-bound and territorial as juveniles in freshwater, becoming pelagic and schooling as they move to sea and mature (Elson, 1957; Hoar, 1958, 1976). I deal with this by considering the social behaviour of fish of any age during those periods when they are active and feeding, and thus when encounters with other conspecifics are most likely to occur.

A more challenging problem is that some species show variation in their social organization at different locations, even when other conditions, such as time of day and tides, are similar. This suggests that group structure and social behaviour may be among a species' adaptations to the local ecological setting, and that one may find clues to the evolutionary origins of fish social organization by close study of such species. The relationship between social behaviour and ecology, which has been examined extensively in some birds and mammals (Wilson, 1975; Crook et al., 1976; Clutton-Brock and Harvey, 1977), has remained largely unexplored among fishes.

## 7.1 Single Fishes

Among fish that live for long periods as separate individuals, two distinct categories can be recognized. First are the *solitary* fishes, so-called because they are generally scattered, and whether relatively mobile or stationary, are not usually found close to conspecifics. Second are species in which individuals occupy territories having contiguous borders; thus a local area is subdivided into a *mosaic* of individually

150

occupied territories. Some territorial mosaics established only for breeding were discussed in Chapter 5. Here I consider mosaics of contiguous territories that are occupied over relatively long periods of time, and not just for breeding.


## 7.1.1 Solitary Fishes

### 7.1.1.1 Chaetodontidae

Among the butterflyfishes there is great variety in the size and structure of social units and in space-related behaviour. This variation can even be found within a single species observed at different locations. Thus the most common social organization of a species can best be determined by extended observations of known individuals at several localities. The studies of Reese (1973, 1975), Fricke (1973a), Itzkowitz (1974b), and Ehrlich et al. (1977) provide some examples.

Table 7.1 lists eight chaetodontid species usually found as solitary individuals, together with some ecological and behavioural characteristics of each. Only one of these (*Chaetodon trifascialis*) was consistently territorial. Individuals actively patrolled their territories, centred over colonies of *Acropora*, and vigorously excluded all chaetodontids of their own and other species. This was the only species among several tested by Ehrlich et al. (1977) that actively attacked model chaetodontids presented to them in their territories. It also appears to be the only chaetodontid that feeds exclusively on polyps of living coral (Hiatt and Strasburg, 1960).

*C. rainfordi*, *C. aureofasciatus*, and *C. plebeius* usually moved about singly in home ranges, with no apparent aggressive behaviour, during daytime. Occasionally they were seen in groups of three to five fish. Most chaetodontids are diurnally active, settling into substrate shelters, and often changing colour at night (Randall, 1968; Ehrlich et al., 1977). The space-related behaviour of some species changes markedly at night. For example, *C. aureofasciatus* and *C. plebeius* formed "roosting groups" in individual clumps of coral at Lizard Island (Ehrlich et al., 1977). From 2 to 30 fish settled in the spaces among coral branches and sheltered there overnight. As they moved in at dusk and out again at dawn there were frequent bouts of jostling and head-down aggressive display with fully erect spiny dorsal fin. Aggression was directed at other chaetodontids and at pomacentrids that shared the same coral overnight.

The most common social unit for both *C. auriga* and *C. lineolatus* at Heron and Lizard Islands (both on the Great Barrier Reef) was the single individual moving over large home ranges, with no evidence of territoriality. On Jamaican reefs, *C. capistratus* and *C. striatus* were likewise seen ranging singly over home ranges (Itzkowitz, 1974b).

Seven of the eight species in Table 7.1 are omnivores, feeding on a wide variety of animal and algal species that are picked either directly off the reef or from sandy patches among clumps of live coral. They are all home-ranging and non-territorial. Only *C. trifascialis*, the strictly carnivorous species, actively defended territories, and these were in areas where its coral food was concentrated.

**Table 7.1.** Typical habitat, diet, and space-related behaviour of eight species of Chaetodontidae, commonly found during daytime as single individuals. H, Heron Island; En, Enewetak Atoll; J, Johnston Is.; L, Lizard Is.; Ja, Jamaica; E, Ehrlich et al. (1977); Ho, Hobson (1974); H&S, Hiatt and Strasburg (1960); I, Itzkowitz (1974b); R, Reese (1975); Ra, Randall (1967)

| Species | Locality observed | Habitat | Diet | Space-related behaviour | Authority |
|---|---|---|---|---|---|
| *Chaetodon trifascialis* (*Megaprotodon strigangulus*) | H, En, J, L | Living coral colonies (*Acropora* spp.) | Coral polyps | Strongly territorial | R, E, H&S |
| *C. rainfordi* *C. aureofasciatus* *C. plebeius* | H, L | Mainly *Acropora* | Omnivorous, mainly coral polyps | Home ranging. Nocturnal "roosting groups" in *C. a.* and *C. p.* | R, E |
| *C. auriga* *C. lineolatus* | H, L | Sand and dead coral rubble | Omnivorous, especially polychaetes and algae | Large home ranges | R, E, H&S, Ho |
| *C. capistratus* *C. striatus* | Ja | Live coral | Omnivorous, especially polychaetes and sea anemones | Home ranging | I, Ra |

## 7.1.1.2 Esocidae

The circumpolar northern pike (*Esox lucius*) and the muskellunge (*E. masquinongy*) of eastern North America are considered to be solitary, relatively sedentary, opportunistic carnivores (Scott and Crossman, 1973; Nursall, 1973). Clear evidence on their space-related behaviour and social structure is difficult to obtain because of their habitat. Typically both occur in relatively warm, slow-moving, heavily vegetated rivers, streams and canals, and in the shallow, weedy bays of warm-water lakes. Some data from tracking of individual fish carrying ultrasonic transmitters have provided information on movements of both species, but the interpretation of these data is not easy.

In a eutrophic, northern Alberta Lake, tagged *E. lucius* moved over wide areas, but generally stayed in shallow water near shore (Diana et al., 1977). Individuals stayed near one location for up to nine days, and then moved elsewhere. Even though some were found at the same locations after being away for several days or weeks, there was no good evidence for well-defined home ranges. Malinin (1969) reported that tagged European *E. lucius* stayed in areas ranging from 50 to 150 m diameter, in broad, open habitat with level substrate, but those released into narrow, rocky-bottom sections of a river moved up to 500 m along the river.

Somewhat firmer evidence comes from a study of *E. masquinongy* carrying sonic tags in a small, shallow lake and stream in southern Ontario (Crossman, 1977). Fish in the lake tended to stay in roughly circular areas of about 300 m diameter, while those in the creek moved further in a linear direction. One fish was found at its initial tag and release site in the lake 15 months after tagging. Six months later it was followed during a 6.4 km spawning migration to the head of the creek; in less than three weeks it was back at the previous lake location. Crossman (1977) concluded that these fish have "...at least an ill-defined home range" (p. 156).

None of these studies provided any direct evidence of non-reproductive social interactions. The preferred habitat of adult esocids makes observation by divers difficult, and indirect evidence on social behaviour, perhaps from radio rather than sonic tags (Crossman, 1977), may be the best obtainable with present technology. The available evidence supports the suggestion that adult esocids are solitary carnivores. This is probably an efficient social system for large species (up to 120 cm length for *E. masquinongy* and somewhat smaller for *E. lucius* (Scott and Crossman, 1973)), that hunt by ambush, lurking among weeds in shallow water, and leaping out at passing prey.

### 7.1.1.3 Miscellaneous

Many other fishes, especially among the fauna of coral reefs, are more or less solitary. Although often difficult to observe because they are either wide-ranging or secretive by day, or only nocturnally active, some information is available from survey studies. For example, based on an intensive examination of Jamaican reefs Itzkowitz (1974b) classified several species as isolates, i.e., solitary and non-territorial. These included two very different types of fishes: fast-swimming, wide-ranging carnivores such as the labrids *Bodianus rufus, Halichoeres poeyi*, and *H. radiatus*, and the soapfish *Rypticus saponaceus*; and slower-moving, more sedentary benthivores, including the filefish *Cantherhines pullus* and the porcupinefish *Diodon hystrix*. Species on Hawaiian reefs classed as solitary by Hobson (1974) include the moray eels (Muraenidae), that live in crevices and caves on the reef, and a variety of slow-moving species such as the trumpetfish *Aulostomus chinensis*, the balloonfishes *Arothron hispidus* and *A. meleagris*, the sharpbacked puffers *Canthigaster amboinensis* and *C. jactator*, and the spiny puffers *Diodon holocanthus* and *D. hystrix*.

None of the above species appears to be territorial, except for the moray eels, that aggressively defend their shelters. The others may be home-ranging, but there is little clear evidence on their space-related behaviour. They are all either predators or omnivores.

*Comments.* The solitary fishes discussed above are mostly home-ranging; only a few are territorial. Some, such as the esocids, move over large areas but return periodically to specific locations. These may be familiar core areas that are used extensively because of optimal feeding conditions. The only solitary chaetodontid known to be strongly territorial (*C. trifascialis*) is also the only one feeding exclusively on coral polyps. This suggests that the food resource is highly valued, and perhaps also relatively scarce, in comparison with the algae and the wide variety of invertebrate organisms eaten by other chaetodontids.

Another important resource protected by a territory or readily available in a small, well-known home range is shelter from predators. The lack of clearly defended territories and discrete home ranges among many solitary fishes may be related to the presence of effective alternative anti-predator adaptations. The pike and muskellunge are large, strong fishes; it is doubtful that as adults they have any effective predator other than man. The same is probably true for the larger moray eels. The slow-moving marine species listed above have alternative defences, such as large spines, or the ability to inflate themselves temporarily by swallowing water. Also, some of them are extremely secretive, relying on ambush techniques to capture their own prey. Finally, some solitary fishes are nocturnally active. In all these cases the value of a territory or home range to counter predation is probably minimal.

## 7.1.2 Territorial Mosaics

The social system by which a local area is subdivided into a mosaic of contiguous territories, each occupied for long periods by a single animal, will be illustrated with examples from two marine families (Blenniidae and Pomacentridae) and one predominantly freshwater family (Salmonidae).

### 7.1.2.1 Blenniidae

The blennies are small, benthic fishes of relatively shallow waters; some of them are strongly territorial throughout their post-larval life. For example, the redlip blenny (*Ophioblennius atlanticus*) is a Caribbean species in which males and females maintain separate but contiguous territories on rocky substrate (Itzkowitz, 1974b; Nursall, 1977). Each resident actively patrols its territory, seldom rising more than a few centimetres above the bottom, and using crevices and holes in the substrate as shelter. A few locations are consistently used as resting areas and lookouts, from which the fish dives for cover when danger threatens. Similar behaviour is shown by the striped blenny (*Chasmodes bosquianus*), that uses empty oyster shells and other cavities as shelters (Phillips, 1971), and the mussel blenny (*Hypsoblennius jenkinsi*), whose major refuges in its subtidal territories are pholadid clam burrows and *Serpulorbis* snail tubes (Stephens et al., 1970).

All three species are highly aggressive in defence of their territories. Their agonistic repertoires include overt attacks, nips and chases, and also a variety of displays. These are either directed frontally with the mouth open, or laterally. Body quivering, head-snaps and distinctive colour changes often accompany the displays. Resident *O. atlanticus* sometimes "flutter", i.e., swim in tight circles around an intruder, and "lie across", i.e., settle down on top of the intruder (Nursall, 1977).

On Barbados reefs young *O. atlanticus* have difficulty establishing themselves among adult territories. A juvenile may be able to hold a small "interstitial territory" (Fig. 7.1) and gradually enlarge this by taking over space from its neighbours. The vigorous defence of space in *O. atlanticus* was illustrated by experiments in which tagged fish from other locations were released onto the study area. Only one of 10 such fish was able to establish a territory among the residents. When territory occupants were removed in another experiment, the nearest neighbours expanded their territories to take over the vacated one (Nursall, 1977).

154

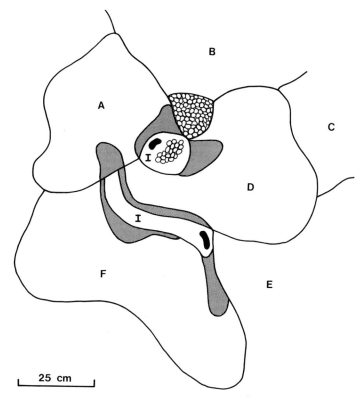

**Fig. 7.1.** Two interstitial territories (*I*) of young *Ophioblennius atlanticus* among territories of adults *A-F*. *Solid black*, crevice retreats of interstitial fish; *stippling*, areas where interstitial fish regularly feed; *clusters of open circles*, patches of live coral. (After Nursall, 1977)

The primary function of territoriality in these blennies is probably the acquisition of space around one or more shelters in which to escape from predators. They are small, substrate-bound fishes and their only effective escape mechanism seems to be to dive into a small hole where a predator cannot follow. *O. atlanticus* feeds on filamentous algae; the other two species are omnivores. Thus in areas where food is abundant, the territorial mosaic social system allows each fish to have sole occupancy over a foraging area around suitable shelter sites.

### 7.1.2.2 Pomacentridae

Many damselfishes distribute themselves over shallow reefs in territorial mosaics. Single adults of both sexes occupy and defend relatively small areas of reef substrate for long periods, and use them for all their major biological activities: feeding, resting, shelter from predation, and breeding (Table 7.2). In addition, these species share a number of other characteristics. They are diurnally active, benthivorous and substrate-bound, rarely rising more than a few centimetres off the bottom, and then only briefly. Except for *H. rubicunda* they are coral reef species, and their territories usually include open sandy areas and patches of coral

**Table 7.2.** Some characteristics of pomacentrid species in which single adults of both sexes maintain long-term territories in a mosaic pattern. H&S, Hiatt and Strasburg, 1960

| Species | Locality | Habitat | Diet | Authority |
|---------|----------|---------|------|-----------|
| *Pomacentrus flavicauda* | Heron Island, G.B.R. | Coral rubble, sand | Mainly algae | Low, 1971 |
| *P. wardi, P. apicalis, Abudefduf lachrymatus* | Heron Island | Coral rubble | Mainly algae | Sale, 1974, 1975; H&S |
| *P. jenkinsi* | Hawaii | Coral rubble, rock, sand | Omnivorous, mainly algae and detritus | Rasa, 1969; H&S; Hobson, 1974 |
| *P. lividus* | Guam | Coral rubble, sand | Algae | Belk, 1975 |
| *P. lividus* | Red Sea | Coral rubble | Algae | Vine, 1974 |
| *P. trilineatus, P. tripunctatus, A. annulatus, A. biocellatus, A. leucozona, A. lachrymatus* | Red Sea | Coral rubble, sand | Mainly herbivorous, algae | Fricke, 1975b; H&S |
| *Eupomacentrus leucostictus* | Caribbean | Coral rubble, sand | Omnivorous, benthos | Brockmann, 1973; Ebersole, 1977 |
| *Hypsypops rubicunda* | California | Rock | Carnivorous, benthos | Clarke, 1970, 1971 |

rubble (i.e., blocks of mostly dead coral, with colonies of live coral around the edges). The territories are remarkable stable. Recognizable individuals of several species are known to have occupied the same locations for many months (Rasa, 1969; Low, 1971; Sale, 1974, 1975). The Californian species *H. rubicunda* may hold the same territories for at least four years (Clarke, 1970). The mosaic character of three sympatric species at Heron Island is illustrated in Figure 7.2.

These maps illustrate the long-term stability and non-overlapping character of territories held by the same species, although slight overlap between heterospecific territories sometimes occurs. Sale (1974) found that virtually all the rubble areas were occupied by members of the three species; areas without territories were regions of extensive live coral, unsuitable for pomacentrid territories, largely because of limited shelter space. Sale also monitored changes in territory occupancy. When *P. apicalis* numbers 2 and 4 disappeared after May 1972, the previous neighbours (especially fish 3) had enlarged their territories to include some of the vacant space by September, when the area was next examined (Fig. 7.2). Juveniles were observed to settle in areas of suitable habitat and gradually enlarge their zone of occupancy as they grew in size. These new sites were usually at the boundaries of existing adult territories, or were centred around shelters that were too small for the adults to use.

Protection of a limited food resource from competitors is a prime function of territoriality in these fishes. The best evidence for this is the vigor of attacks against intruders that share the same food. A clear example is *P. flavicauda*, primarily an algal grazer, that attacks herbivores and omnivores of many species, but does not attack carnivores (Low, 1971). When Vine (1974) moved algae-covered blocks of

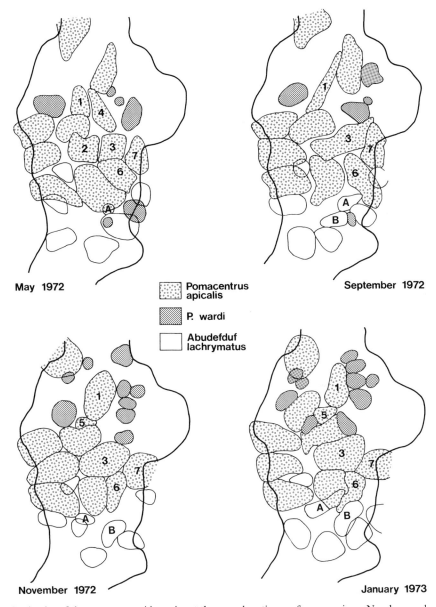

**Fig. 7.2.** Territories of three pomacentrid species at the same location on four occasions. Numbers and letters refer to known individuals. (After Sale, 1974)

rubble outside the territory of a *P. lividus*, other fishes quickly descended on the displaced blocks and ate the thick algal turf within 30 min. Clearly the resident fish had been protecting a resource that was in demand. The central role of food protection in territoriality of omnivorous pomacentrids was demonstrated quantitatively by Ebersole (1977) who found a positive relationship between the level of aggression directed at potential intruders by territory-guarding *E.*

157

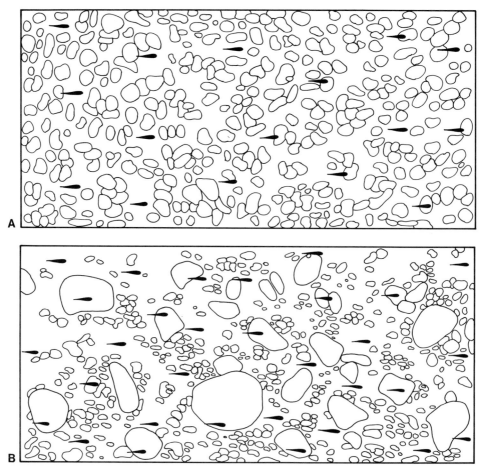

**Fig. 7.3.** Variation in number of *S. salar* underyearlings able to maintain territories on gravel substrate with (**B**) and without (**A**) large boulders present. (After Kalleberg, 1958)

*leucostictus* and the Potential Competitive Impact (PCI) of intruder species. PCI is an expression combining the degree of diet overlap between intruder species and *E. leucostictus* with the intruder's relative rate of food consumption (based on size).

Thus, in general it seems clear that the territorial mosaic system found in many pomacentrids is a behavioural mechanism for distributing resources of food and shelter among a local population. The strong inter- and intraspecific territoriality shown by many species indicates that competition for food is vigorous on coral reefs, and territory-holders must expend much energy to retain their local source of food.

### 7.1.2.3 Salmonidae

The territorial mosaic social system is common among juveniles of salmonid species that spend the first months or years of their lives in rivers and streams. Examples are Atlantic salmon (*Salmo salar*) and coho salmon (*Oncorhynchus kisutch*). Lake-

**Table 7.3.** Number of territory-holding *S. salar* underyearlings on successive days as the water current speed was raised and then lowered. (From Kalleberg, 1958). Differences between mean at high speed and at both lower speeds significant (*t*-test, $p < 0.01$)

| Current speed (cm/s) | 18 | | | | | 29 | | | | | 17 | | | | |
|---|---|---|---|---|---|---|---|---|---|---|---|---|---|---|---|
| Date in July, 1954 | 12 | 13 | 14 | 15 | 16 | 17 | 18 | 19 | 20 | 21 | 22 | 23 | 24 | 25 | 26 |
| Number of territories | 244 | 259 | 264 | 236 | 253 | 280 | 298 | 289 | 261 | 279 | 243 | 262 | 234 | 247 | 255 |
| Mean | 251.2 | | | | | 281.4 | | | | | 248.2 | | | | |

dwelling juvenile salmonids (e.g., sockeye salmon, *O. nerka*) and those that move directly to sea after emerging from the gravel where they hatched (e.g., pink salmon, *O. gorbuscha*), typically live in schools.

Stream-living territorial salmonids usually rest on the substrate, facing upstream. A resident fish remains mostly at a particular location or "station" within its territory. From there it darts away to pick food off the bottom or from the drift, or to challenge another fish approaching its territory, and then returns to the "station" (Kalleberg, 1958; Keenleyside, 1962).

The size of individual salmonid territories varies with the degree of visual isolation from neighbouring territory-holders, current speed and prey density. Figure 7.3 shows the difference in number of underyearling *S. salar* territories within mosaics on gravel substrate with and without large boulders. The visual blocks provided by the boulders allow some fish to approach their neighbours more closely than on a smooth substrate before eliciting aggression.

Velocity of stream flow influences territory size because as flow decreases the fish are less substrate-bound; they can more easily hold position at the territory station by active swimming in open water (Hartman, 1963). As they move up off the bottom, visual contacts with nearest neighbours increase and mean territory size goes up. Documentation of this effect in young *S. salar* is shown in Table 7.3.

Finally, territory size tends to vary inversely with prey density. This relationship was well demonstrated in a study by Slaney and Northcote (1974), in which *S. gairdneri* underyearling territories in laboratory stream channels increased in size as the amount of food (prepared hatchery food and stream drift organisms) declined (Fig. 7.4). As prey quantity decreased, the fish made longer excursions searching for food and their territory boundaries expanded.

The territorial mosaic seems to be a common social system among salmonids when all the fish of a local population are similar in size, for example, shortly after the underyearlings in one area emerge from the gravel and begin their active, feeding existence while remaining close to the substrate. The gravel remains the source of most food and of shelter, and since the former is not likely to be uniformly distributed, some individuals will grow faster than others. Faster-growing fish will establish themselves where food and shelter are most available; slower-growing fish are kept out of these areas, and relegated to peripheral regions, where conditions may not be as good (Mason and Chapman, 1965; Reimers, 1968; Slaney and

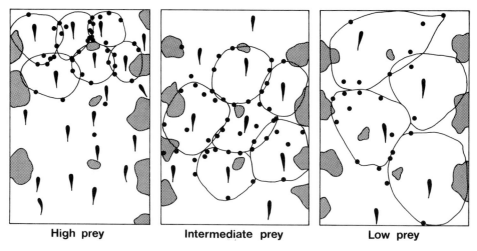

**High prey**   **Intermediate prey**   **Low prey**

**Fig. 7.4.** Effect of prey abundance on size of *S. gairdneri* underyearling territories. *Solid circles*, location of aggressive interactions; stippling, stones. (Modified after Slaney and Northcote, 1974)

Northcote, 1974). Thus the mosaic of similar-sized, contiguous, more or less equivalent territories may gradually be replaced by a system in which larger fishes are dominant over smaller ones, and have readier access to regions of the stream where food and shelter are optimal. This situation is most clearly seen in populations involving fish of various ages and sizes. Examples are *S. gairdneri, S. trutta*, and *Salvelinus fontinalis*, where adults and juveniles of various ages often share the same regions of a stream. In such situations the territorial mosaic is replaced by a combination of territoriality and dominance ranking (Newman, 1956; Jenkins, 1969).

The biological significance of territoriality among stream-living salmonids has been considered by many workers. Most have argued that the securing of food resources and of shelters in which to avoid predators are the primary funcitions. The diet of these species consists of organisms picked off the substrate or from stream drift. In either case, securing an area of substrate free of potential competitors is seen as an advantage. In addition, prolonged residence on a small area of stream bottom must increase familiarity with substrate topography, so that the fish can quickly dive for shelter when predators approach. Despite the great interest in ecology and behaviour of salmonid fishes there is surprisingly little experimental evidence bearing directly on the function of stream territories. The following two examples show that artificial streams can be useful in studying this problem.

Slaney and Northcote (1974) varied the amount of food available to *S. gairdneri* underyearlings in laboratory stream channels and measured the effects on territorial behaviour. The results were clear-cut; fish density increased as prey abundance increased. This was associated with a decrease in territory size (Fig. 7.4). As prey levels rose the fish were apparently able to secure adequate food within smaller defended areas. Further there was an inverse relationship between prey density and the frequency of aggressive behaviour, in experiments where the density

160

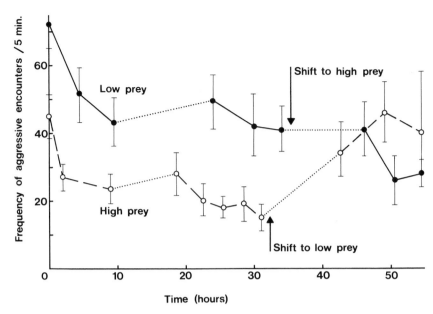

**Fig. 7.5.** Effect of prey abundance on aggressive encounter frequency in *S. gairdneri* underyearlings held in laboratory channels. *Arrows* show when prey abundance was reversed. *Vertical lines*, 95% confidence limits. (Modified after Slaney and Northcote, 1974)

of underyearlings was held constant (Fig. 7.5). This shows that decline in territory size as food abundance rose was a direct result of less aggression by the fish. This was emphasized by a dramatic reversal of aggression levels when prey abundance was reversed during the experiment (Fig. 7.5).

Symons (1974) showed that the opportunity to establish territories in a large artificial stream tank gave young *S. salar* (7 to 10 cm long) an advantage in avoiding predation by large (34–39 cm) brook trout (*Salvelinus fontinalis*). Groups of 15 salmon that had been in the stream for 15 days before addition of predators (for two days only) suffered less mortality than groups with only one day's experience in the stream. Losses were 5 and 15 salmon respectively over four replications of the experiment. Since the young fish soon established territories in the tank, which had gravel and large boulders as substrate, it seemed clear that longer possession of a territory allowed the fish to become more familiar with the surrounding substrate. This provided the more experienced salmon with greater opportunities to avoid capture by the trout.

*Comments.* The blenniids, pomacentrids, and young salmonids discussed above are all relatively small, substrate-bound fishes living in shallow water. The mosaic of contiguous territories they occupy is basically a flat, two-dimensional system, and the territory occupants obtain their essential resources (food, resting sites, shelter and, in the case of adults, breeding sites) from the substrate, or from the water column close to the bottom. They are also diurnally active fishes, remaining inactive overnight in burrows, holes or crevices in the territory (blennies and

161

damselfishes) or in deeper pools or very shallow, shoreline areas (young salmonids).

The mosaics are for the most part occupied by fishes of about the same size. The young salmonids that occupy adjacent territories are usually from the same year-class, and if differing growth rates produce variation in size among a local population, the different size-groups usually separate from each other, the larger fish moving into deeper water. Among the marine species, juveniles have difficulty establishing territories within a mosaic, and usually must wait in small, interstitial spaces among the existing territories for vacancies to appear through death or desertion of residents before they can move in and maintain a long-term territory.

The occupancy of discrete territories by single fish rather than by groups appears to be related to the requirements of food and shelter. Long-term utilization of benthic food can be assured by the mosaic system, each individual guarding its own resources. At the same time, thorough familiarity with escape routes and shelters within the territory reduces vulnerability to predators. Exclusive occupancy of a small, well-known territory thus assures the resident fish of ready access to both food and shelter.


## 7.2 Male–Female Pairs


The social unit of a bonded pair, in which one male and one female remain together for long periods, and breed repeatedly, is rare among fishes. Pair bonds among biparental cichlids last up to several weeks, as the parents collaborate in raising a brood until the young fish disperse. In aquaria, without alternative breeding partners, single cichlid pairs may breed repeatedly, but unequivocal evidence from free-living fishes of extended pair formation, lasting beyond a single breeding episode, is rare. The best available examples are among reef-dwelling chaetodontids.

Table 7.4 lists 12 butterflyfishes in the genus *Chaetodon* that have been observed living in single male–female pairs for periods up to 12 months (Reese, 1973). In the Red Sea some pairs of tagged chaetodontids have been seen together for 3 years (Fricke, 1973a). Eight species in the table are classed as strongly paired, four as weakly paired. The distinction is somewhat arbitrary, and may disappear with further evidence, but is based on the frequency with which the species was seen in pairs, as compared with all observations of that species.

Strong evidence for pair bonding is provided by the distinctive behaviour of the members of a pair that have become separated while feeding on the reef. One fish rises a little off the bottom, rolls or tilts sideways while turning, as though looking for its partner. On seeing each other the fish quickly reunite, and swim about close together in an agitated manner near the bottom, in what has been termed a "greeting ceremony" by Reese (1975). The pair soon settles down and resumes feeding together on the reef. Ehrlich et al. (1977) saw similar behaviour and suggest the tilting behaviour of the separated fish may serve a signal function, since sunlight often reflects off the white, yellow and black "poster colour" patterns of most

Table 7.4. Some characteristics of 12 species of Chaetodontidae that live primarily in single, male–female pairs. En, Enewetak Atoll; H, Heron Island; Ha, Hawaii; L, Lizard Island; J, Johnston Island; E, Ehrlich et al. (1977); R5, Reese (1975); R7, Reese (1977); H&S, Hiatt and Strasburg (1960); Ho, Hobson (1974)

| Species | Locality observed | Paired | | "Greeting ceremony" | Diet | Space-related behaviour | Authority |
|---|---|---|---|---|---|---|---|
| | | Strongly | Weakly | | | | |
| *Chaetodon triangulum* | H, L | + | | + | Coral | Territorial | R5, R7, E |
| *C. multicinctus* | H | + | | | Coral | Home range or wandering | R5, R7, Ho |
| *C. punctato-fasciatus* | En | + | | | Coral | Home range or wandering | R5, R7 |
| *C. tri-fasciatus* | H, En, Ha, L | + | | + | Coral | Home range, nocturnal "roosting" group | R5, R7, E |
| *C. ornatissimus* | H, En, Ha | + | | | Coral and coral mucous | Home range, maybe territorial | R5, R7, Ho |
| *C. ephippium* | H, En, J, L | + | | + | Coral and algae | Home range or wandering | R5, E, H&S |
| *C. unimaculatus* | H, En, Ha, L | + | | + | Omnivorous with coral | Home range or wandering | R5, R7, E, Ho |
| *C. vagabundus* | H, L | + | | | Coral and algae | Home range or wandering | R5, R7, E, H&S |
| *C. citrinellus* | E, L | | + | + | Omnivorous with coral | Home range | R5, R7, E, H&S |
| *C. lunula* | E, Ha, L | | + | | Omnivorous with coral. | Home range | R5, E, Ho |
| *C. quadri-maculatus* | Ha, J | | + | | Coral and polychaetes | Home range | R5, R7, Ho |
| *C. reti-culatus* | E, Ha | | + | | Coral | Home range | R5, R7 |

chaetodontids, making the tilting fish very conspicuous. Species seen performing "greeting ceremonies" are indicated on Table 7.4.

All 12 species feed on live coral and some also ingest other benthos, including algae. The first five species in Table 7.4 are classed as obligate coral feeders by Reese (1977), and one of these (*C. ornatissimus*) may be unique among chaetodontids in sucking mucous from living coral rather than biting off the polyp tips as the other species do (Hobson, 1974). *C. triangulum* is the only species in the table that is consistently territorial, although *C. ornatissimus* pairs occasionally are aggressive to each other. The others are home ranging, and some are also classed as occasionally wandering (i.e., not consistently associated with any area of the reef). Clear distinction between these two categories probably requires more observational records than are now available. Most chaetodontids are typically active by day, but *C. lunula* is apparently an exception. In Hawaiian waters it feeds mainly at night and during the day gathers in small groups or large schools, but does not feed (Hobson, 1974).

The functional significance of the long-term male–female pair as a social unit in chaetodontids is open to speculation. It may be an efficient system for species that range widely while feeding mainly on coral. Since many coral species only fully extend their polyps at night (Campbell, 1976), the search for available polyps by day may require wider ranging than does the search for other benthos. If their diet requires a wide-ranging search strategy, then close association in male–female pairs will simplify periodic breeding. Little is known in detail about chaetodontid reproduction, but they apparently breed in single pairs and release pelagic eggs (Breder and Rosen, 1966; Reese, 1975). Remaining in firmly bonded pairs for long periods will allow these species to breed at intervals without the need to divert energy into time-consuming territory maintenance, mate selection, and extended courtship. Further study of the strongly territorial species, *C. triangulum*, may be useful in testing this hypothesis.

## 7.3 Small Groups

### 7.3.1 Anemonefishes (Pomacentridae)

Pomacentrids of the genera *Amphiprion* and *Premnas* are the well-known tropical anemonefishes or clownfishes, that live symbiotically with large sea anemones (Mariscal, 1972). Several *Amphiprion* species have been closely studied by ethologists and knowledge of their social and space-related behaviour is rapidly accumulating. The most common social unit is a monogamous adult pair, together with several subadults, associated with a single anemone (Mariscal, 1972; Brown et al., 1973; Moyer and Sawyers, 1973; Fricke, 1974; Allen, 1975; Ross, 1978). The adults are strongly territorial, actively driving other fishes away from their home anemone. The subadults may have small sub-territories within the larger territory of the pair (Graefe, 1964). Within the group the adult female is larger than and dominant over the male, and both adults are dominant over the smaller subadults (Allen, 1975). The anemone is the focus of all the group's activities: feeding, resting, sheltering from predators, and breeding. Individuals rarely move more than a short distance from their host anemone, and then only to feed on plankton which, together with benthic algae, makes up most of their diet (Fishelson et al., 1974; Allen, 1975). The interrelationships between anemonefishes and their hosts, including arguments about the benefits derived by both members of the symbiosis, were reviewed by Mariscal (1972).

The pair bond between male and female *Amphiprion* is durable, lasting in some cases for three years (Fricke and Fricke, 1977). Experiments in their natural habitat showed that *A. bicinctus* recognize their mates when presented with several alternatives in a choice apparatus (Fricke, 1973a). Visual cues seem to be primary in this recognition; changing a fish's appearance with dye or a plastic covering led to immediate attack by its partner.

Even more exciting was the discovery that *A. bicinctus* and *A. akallopisos* are protandric hermaphrodites (Fricke and Fricke, 1977). All individuals are first males; those surviving until they are able to replace the female in a social group, become female. The monogamous mating system is basically maintained by the male of the adult pair. He is socially subordinate to his larger mate, but is larger than and dominant over all subadults living with the same anemone. He alone breeds with the female, driving smaller males away if they attempt to breed. If the female disappears, her mate immediately begins transforming into a female, and the largest of the associated subadults becomes the new male breeding partner. The new female can begin laying fertile eggs within 26 days of changing sex. If both adults are removed the subadults rapidly increase in size, with two of them soon forming a new adult pair (Allen, 1975). Similar sex change by the male of an established pair following loss of the female was documented for *A. melanopus* at Guam (Ross, 1978), and this trait may well turn out to be widespread within the genus *Amphiprion* as more information accumulates (Moyer and Nakazono, 1978).

The social unit of a monogamous adult pair with several subadults seems to be an efficient system for fishes living in apparently obligatory symbiosis with sea anemones. In general, the host organisms are unpredictably distributed and not numerous at one location (although colonies do occur, see below). Once an *Amphiprion* pair is established at an anemone they breed repeatedly, while gaining shelter from their host, and feeding in its immediate vicinity. The intensity and vigor of their territorial aggression suggests there is a high premium on remaining with the host. The subadults provide the reservoir from which new adults can be recruited quickly if one or both adults disappear (Fricke and Fricke, 1977).

However, there is some flexibility in the system. In Japanese waters *A. clarkii* were found in monogamous pairs with subadults, in locations where host anemones were scattered (Moyer and Sawyers, 1973). However, where the anemones occurred in large colonies, single female *A. clarkii* often had a large territory that included the smaller territories of two or three males plus subadults. Each male dedended a small core area that contained a nesting site, and most of his aggression was directed at neighbouring males. The female appeared less overtly aggressive, except in joint defence with a male of his nest site, when eggs were present there (Moyer, 1976; Moyer and Bell, 1976). Fricke (1974) also suspected that some *A. bicinctus* females in the Red Sea were polyandrous, with a single large female visiting and breeding with more than one nest-guarding male.

Polyandry can be expected in populations where protandry is the rule and adult females are dominant over adult males. Since all the subadults are male, then if many hosts are available, a female can increase her fitness by acquiring a large territory containing more than one breeding site with a mate. Since adult males without mates can soon become females, the polyandrous female can restrict the number of females competing with her for nest sites by remaining dominant over more than one male within her single large territory.

## 7.3.2 Dascyllus (Pomacentridae)

Several species of the coral reef genus *Dascyllus* are typically distributed in small, stable groups of two or more adults and usually several subadults. Each group is closely associated with an individual colony of a branching type of coral, and is

**Table 7.5.** Group sizes of *Dascyllus* species

| Species | Locality observed | Group size | | | Authority |
|---|---|---|---|---|---|
| | | Mean | Typical Range | Maximum | |
| *D. aruanus* | Heron Island | | 2–5 | 25 | Sale, 1972 |
| *D. aruanus* | Kenya | 20.9 | — | — | Brown et al., 1973 |
| *D. aruanus* | Red Sea | — | 3–6 | 14 | Fricke and Holzberg, 1974 |
| *D. aruanus* | Red Sea | — | 5–30 | — | Fishelson, 1964 |
| *D. aruanus* | Red Sea | — | 10–20 | — | Fishelson et al., 1974 |
| *D. reticulatus* | Kenya | 7.9 | — | — | Brown et al., 1973 |
| *D. marginatus* | Red Sea | — | 5–12 | 60 | Holzberg, 1973 |
| *D. marginatus* | Red Sea | — | 10–20 | — | Fishelson et al., 1974 |
| *D. trimaculatus* | Kenya | 7.1 | — | — | Brown et al., 1973 |
| *D. trimaculatus* | Red Sea | — | — | 80 (adults only) | Fricke, 1973c |
| *D. trimaculatus* | Enewetak | — | — | 60 (adults only) | Allen, 1975 |

territorial, defending the home coral against intrusion by fish of the same and other species. The association is persistent; marked *D. aruanus* at Heron Island were found at the same coral block for up to seven months (Sale, 1971b), and tagged *D. trimaculatus* occupied the same corals in the Red Sea for up to two years (Fishelson et al., 1974). The fish feed in the open water around the home coral, and use the interstices among the coral branches for shelter and for resting places at night. The diet is omnivorous, consisting mainly of plankton (Fricke, 1973c; Fishelson et al., 1974), although *D. aruanus* also crops filamentous algae from the surrounding coral (Hiatt and Strasburg, 1960).

Group size varies among species and among localities within a species (Table 7.5). This is especially marked in *D. aruanus*. In the Red Sea Fricke and Holzberg (1974) found that 38% of the groups of *D. aruanus* they saw were single male–female pairs, while the others ranged from 3 to 14 fish. Group size is determined to some degree by the size of the home coral block. At Heron Island the number of *D. aruanus* in a group varied directly with home coral size, in regions where that species of coral was scarce and was being fully utilized by *D. aruanus* (Sale, 1972). In the Red Sea the size of host coral blocks also seemed to determine the group size in *D. aruanus* (Fricke and Holzberg, 1974) and *D. marginatus* (Holzberg, 1973).

*D. trimaculatus* is more variable in group structure and association with the substrate than the other studied species. Juveniles are often found living near or among the tentacles of sea anemones (Brown et al., 1973; Fricke, 1973c; Allen, 1975) and the spines of large sea urchins (Fricke, 1973c). In some locations aggregations of 60 or more non-breeding adults have been observed feeding in open water above the reef (Fricke, 1973c; Allen, 1975). Similar aggregations of *D. marginatus* have been seen in the Red Sea (Holzberg, 1973).

166

In *D. aruanus* the groups consisting of more than a single pair appeared to be polygynous harems, with a single dominant male (70–80 mm length) and several females ranging from 20 to 70 mm (Fricke and Holzberg, 1974). The females maintained a rank order by size, and when breeding occurred they spawned in sequence according to rank, the largest first. Examination of the gonads of 11 fish in one group showed there was one adult male, eight females, and two fish with gonadal tissue of both sexes. Since the dominant male is typically more aggressive towards male than female conspecifics approaching the group, Fricke and Holzberg suggest that roving subadults have easier access to an established group if they are female or hermaphroditic. Once in the group, if they become functional females they can enter the female hierarchy and occasionally spawn. As they grow in size they rise in rank and presumably spawn more frequently.

The *Dascyllus* social system, with a strongly territorial small group associated for long periods with a single coral colony, is well suited for coral species that have many small branches, and can thus provide hiding places for fish of different sizes. All the basic requirements, such as feeding, shelter, resting area at night or in turbulent water, and breeding can be met by the single coral colony.

## 7.4 Schools

The final social unit considered in this chapter is the fish school. This is an aggregation within which there is no apparent behavioural differentiation among members. There are no leaders, no persistent ties among near-neighbours, no pair bonds and no dominance relationships. There appears to be no overt aggression, although the regularity of spacing between individuals, especially in fastmoving, well-coordinated polarized schools (see below) strongly suggests that a typical individual distance is maintained. All members of a school tend to perform the same activity together. It is the behavioural equivalence of all members and the synchrony of their ongoing behaviour that puts the large fish school into the category of a mass grouping phenomenon, together with many large bird flocks, mammal herds, and insect swarms.

Schooling behaviour is widespread among fishes (Shaw, 1978). It occurs in some species in virtually all aquatic habitats, and at all taxonomic levels, from primitive to advanced. There is, however, great variation in the persistence of the behaviour. Some fishes swim in schools from the beginning of independent locomotion until death, including periods of breeding. Members of the Clupeidae, Atherinidae, Scombridae, and Cyprinidae are commonly cited examples. Others school only during certain stages of their life history. For example, coho salmon (*Oncorhynchus kisutch*) and Atlantic salmon (*Salmo salar*) spend the early part of their lives as territorial, stream-living fishes; then they form schools, migrate to sea, and remain schooling until they spawn. Some cichlids live in school as fry, gradually dispersing as they develop.

Even more striking is the variation in intensity of schooling behaviour. This has led to a distinction between *polarized* and *non-polarized* schools (Shaw, 1970). In the former, all fish swim parallel to their neighbours, maintain consistent spacing among themselves, and change speed and direction in unison. In general the whole group moves as though it was a single, large organism. Non-polarized schools are more diverse in internal structure and behaviour. Spacing, orientation, and speed of swimming tend to be variable. The entire group is often stationary or only slowly moving. An interesting extreme case occurs when the individual distance decreases to virtually zero, as all members of the group push together into a mass. These contact schools may be flat and two-dimensional as in *Mugil cephalus* (where they are referred to as "pods"), or spherical, as seen occasionally in young *Sebastodes paucispinis*. Photographs of both cases are shown in Breder (1959). The causation and biological significance of such tightly packed masses of fish are not well understood.

Many fishes swim in schools that are alternately polarized and non-polarized, depending on their activities at the time. For example, the threespine stickleback (*Gasterosteus aculeatus*), when undisturbed, often forms non-polarized feeding schools. The group either remains stationary or slowly moves along, close to the bottom, as each fish forages. Spacing and orientation among the individuals are variable. There appears to be little cohesion or synchrony of movement among the members of the group. However, if alarmed, the same fish will quickly close ranks, form a polarized school, and move away from the source of disturbance as a cohesive unit, the individuals closely and uniformly spaced, parallel to each other and swimming at the same speed (Keenleyside, 1955).

Some authors use the term "aggregation" for non-polarized schools, and "school" only for the polarized units. However, I consider it more useful to reserve "aggregation" for yet another type of fish group, that formed when a number of fish independently respond to a common external stimulus (e.g., a localized food supply, or a single point source of light) by approaching and staying near it. These fish are close together, not because of responses to each other, but because they all respond in the same way to the patchily distributed food, or the spot of light. The key distinction then between a school and an aggregation is that only the former is created by "biosocial attraction among the fish" (Shaw, 1970). These and other labels for the various types of fish groupings have been discussed at length in the literature (e.g., Parr, 1927; Keenleyside, 1955; Breder, 1959, 1967, 1976; Shaw, 1962, 1970; Williams, 1964; Radakov, 1973).

## 7.4.1 Behaviour of Fish in Schools

The most highly polarized schools, providing the classic examples of the phenomenon, are found among pelagic, marine species such as clupeids and scombrids. Because of their mobility, speed and open ocean habitat, close observation and study of such species is difficult. Thus most studies of schooling behaviour in natural habitats have been made in the relatively shallow waters of coastlines, coral reefs, small lakes, and streams. For example, the spottail shiner (*Notropis hudsonius*) is a North American cyprinid often found in large, dense schools. Two main types of locomotor behaviour occur in these schools (Nursall, 1973). Individuals in a stationary school move about slowly in a series of short,

168

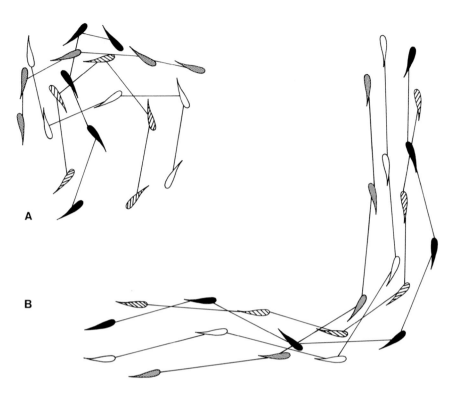

**Fig. 7.6.** Diagrammatic representation of paths followed by four *Notropis hudsonius* showing short radius behaviour (**A**) and loose cruising association (**B**). (Modified after Nursall, 1973)

jerky forward movements, consisting of one or two tail beats, a forward glide, a sharp turn, more tail beats, and so on. This has been called *short radius behaviour* (Fig. 7.6A). The net effect of this activity by all members of a school is that the group remains almost stationary. *N. hudsonius* also swims in a *loose cruising association*, in which individuals move forward more or less parallel to each other, with only slight changes in direction; all fish therefore stay together in a loosely organized group that moves steadily forward (Fig. 7.6B). A school changes from short radius behaviour to the loose cruising association in response to water currents and to mild disturbance. However, if a school is suddenly attacked by a predator, those fish at the point of attack may "explode" in all directions away from the predator, and then move back into the body of the school, in "flash expansion" behaviour (Nursall, 1973).

A special form of schooling behaviour is the *mill*, in which all members of a group follow each other in a closed circle, those at the perimeter moving faster than those near the centre of the circulating mass. Once a mill is formed it often persists until some sudden event, such as approach to a shoreline, or a predator's attack, breaks up the formation. Milling behaviour was first described in detail by Parr (1927) who suggested that it begins when the leading members of a moving school for some reason suddenly turn through about 180°. The leading fish then see those following behind, turn to join them and a closed circle of moving fish is formed.

169

Breder (1951, 1959) has long been interested in the causes of mill formation. He observed, for example, that schools of small mullet (*Mugil cephalus*) in shallow water sometimes shift into the rapidly rotating mill formation with no sudden turning of the lead members. These were referred to as "intrinsic mills". Their causation is not well understood, but the hydrodynamic shearing forces that lead to the formation of vortices in the water surrounding actively moving fish may be involved (Breder, 1965).

### 7.4.1.1 Relatively Stationary Schools

The coral reef environment provides many examples of typical schooling fishes. Among the most conspicuous are those that hover in more or less stationary schools in open water near the reef crest, where it falls away steeply into deeper water. Often these schools are closely associated with large, conspicuous coral heads that project above the reef crest. These species appear to have extremely limited home ranges, remaining closely associated with the same coral head for many months (Popper and Fishelson, 1973; Fishelson et al., 1974). Behaviour of the fish in such schools is synchronized: maintaining position, feeding on plankton, diving for cover into crevices in the reef, and returning from shelter to open water are performed by all fish together. The degree of polarization of fish within the school depends largely on current; polarization increases as current flow increases, and all fish swim actively to maintain their position on the reef.

Some of these schools consist almost entirely of fish of one species. For example, the small pomacentrid *Chromis caeruleus* is often found in large, unispecific schools, within which juveniles tend to stay together in a sub-school (Fishelson et al., 1974). More typically they are multi-species assemblages within which the members of each species tend to stay together, but the feeding and shelter-seeking behaviour of the entire group is well-synchronized. The internal structure and the orientation of individuals within a mixed-species school studied in the Red Sea by Popper and Fishelson (1973) is shown in Figure 7.7. The main component is *Anthias squamipinnis*, with females, males, and juveniles maintaining different positions within the school. A few males in breeding condition hold territories on the large rocky outcrop that is the central focus of the school. Smaller numbers of the pomacentrid *Abudefduf azysron* are typically found in the upper part of *Anthias* schools, and other species occasionally also join the assemblage. Synchronized behaviour within the school is most conspicuous when the current is strong and planktonic feeding is intense.

The influence of current on schooling behaviour was also studied by Potts (1970) at Aldabra in the Indian Ocean. A school of the snapper *Lutjanus monostigma* was observed for many days in a narrow channel between two islands (Fig. 7.8). At slack water between tides the fish formed a dense ball that slowly rose to the surface. As tidal flow began the fish dispersed somewhat into a polarized school, with each fish heading into the current and maintaining position by positive rheotaxis. With increasing current speed they settled to the bottom, and eventually took shelter behind boulders.

Using a small submarine off the island of Oahu, Hawaii, Brock and Chamberlain (1968) found that stationary schools of fish were common over rocky

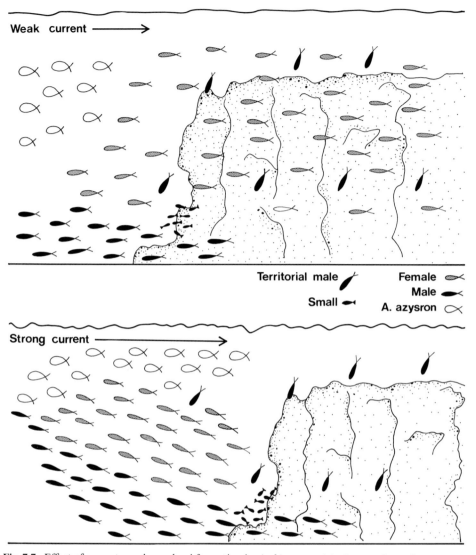

**Fig. 7.7.** Effect of current speed on school formation in *Anthias squamipinnis* around a rocky outcrop. See text for details. (After Popper and Fishelson, 1973)

outcrops and low escarpments at about 70 m depth, and around conspicuous crests above the edge of massive escarpments at about 120 m. The species composition of the schools was different at the two depths, but in both cases they were closely associated with conspicuous outcrops or crests. At the deeper level the schools took the form of tall, narrow columns of fish extending 15 to 40 m above the substrate. The most numerous species was the butterflyfish *Chaetodon miliaris*, and other species often seemed to be associating with the large columns of *C. miliaris* rather than independently with the outcroppings. The explanation of this discontinuous distribution of fishes along the two escarpments is not clear. Most of the species, including *C. miliaris*, were also found closer to shore over living coral. They are

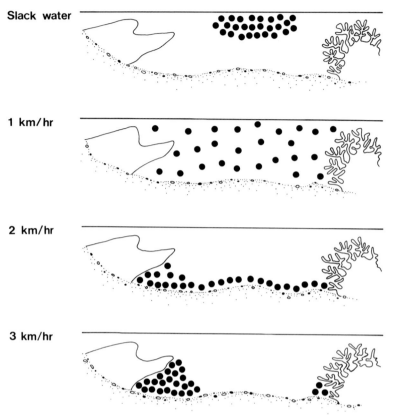

**Fig. 7.8.** Diagrammatic view of the distribution of *Lutjanus monostigma* within a school in relation to current speed. Drawing represents cross-section of a channel between a rock (*left*) and a coral head (*right*). (After Potts, 1970)

diurnally active, planktivorous species, and it has been suggested that deep water plankton that migrate upward at night may be swept inshore over shallower reefs, become trapped there, and thus available to diurnal planktivores (Brock and Chamberlain, 1968). Why the fish should also congregate over outcroppings and reef crests in deeper water is not known. Their numbers are so great that the substrate at those locations is not likely to provide shelter for them all when alarmed. Perhaps the protective value of the stationary school itself is of primary importance, attracting fish of several species and sizes, including those that would not be able to stay together in a single, fast-moving school.

The anti-predator function of large, more or less stationary schools is also suggested by the behaviour of some grunts (Pomadasyidae) and goatfishes (Mullidae) that feed at night as single individuals on reef benthos, but hover over the reef edge by day in large, often mixed-species schools (Randall, 1968; Hobson, 1974). Shelters within the reef structure are generally occupied during daytime by other nocturnal feeders, such as apogonids, holocentrids, and muraenids, and the clustering into large, inactive heterotypic schools may provide an alternative form of protection against predators (Ehrlich and Ehrlich, 1973).

172

**Table 7.6.** Diet and feeding habitat of some schooling surgeonfishes (Acanthuridae)

| Species | Locality studied | Habitat while feeding | Diet | Authorities |
|---|---|---|---|---|
| *Acanthurus guttatus* | Pacific Ocean islands | Reef flat | Algae | Barlow, 1974a |
| *A. triostegus* | Pacific Ocean islands | Reef flat | Algae | Barlow, 1974a, b; Randall, 1961a; Hiatt and Strasburg, 1960 |
| *A. lineatus* | Pacific Ocean islands | Surge channels | Algae | Barlow, 1974a |
| *Zebrasoma veliferum* | Marshall Islands | Reef flat | Algae | Hiatt and Strasburg, 1960 |
| *A. sohal* | Red Sea | Shallow patch reefs | Algae | Vine, 1974 |
| *A. coeruleus* | Caribbean | Shallow patch reefs | Algae | Alevizon, 1976 |
| *A. bahianus* | Jamaica | Sand and grass patches | Algae | Itzkowitz, 1974b |
| *A. gahhm* | Pacific Islands | Sand patches | Algae | Randall, 1956; Hiatt and Strasburg, 1960 |
| *A. dussumieri* | Pacific Islands | Sand patches | Algae, diatoms and detritus | Barlow, 1974a; Jones, 1968 |
| *A. mata* | Pacific Islands | Sand patches | Algae, diatoms and detritus | Barlow, 1974a; Jones, 1968 |
| *Naso literatus* | Marshall Islands | Sand and rock | Algae | Hiatt and Strasburg, 1960 |
| *A. leucopareius* | Pacific Islands | Edge of reef front | Algae | Barlow, 1974a |
| *N. unicornis* | Marshall Islands | Coral heads | Algae | Hiatt and Strasburg, 1960 |
| *A. thompsoni* | Pacific Islands | Off cliffs and pinnacles | Plankton | Barlow, 1974a; Hobson, 1974; Jones, 1968 |
| *N. hexacanthus* | Pacific Islands | Off reef edge | Plankton | Barlow, 1974a; Hobson, 1974; Jones, 1968 |

## 7.4.1.2 Relatively Mobile Schools

Many schooling fishes are more mobile than those discussed above. Open ocean, pelagic species that live in constantly moving, polarized schools are the best examples, but they are notoriously difficult to study in their natural habitat. Shallow, inshore waters provide better opportunities for detailed observation, and here again valuable insights into the relations between social structure and ecology are coming from work on coral reef fishes.

Some of the surgeonfishes (Acanthuridae) are territorial, but many are schooling, and because of their abundance, mobility, and striking colour patterns, are among the most conspicuous components of coral reef communities. Some characteristics of schooling acanthurids, studied at different localities, are listed in Table 7.6. Most of these species are herbivorous, but the two planktivores, *Acanthurus thompsoni* and *Naso hexacanthus*, are found typically in school off the edge of reefs, at various depths down to 100 m or more (Strasburg et al., 1968; Brock and Chamberlain, 1968). The herbivorous species occupy a variety of reef zones, generally in shallow water. Some, such as *A. guttatus* and *A. triostegus* are highly mobile, and move onto the shallow reef flat zones to graze on algae at high tides (Barlow, 1974a). *A. bahianus* often remain over coral blocks as single individuals, then join to form groups that move out over open grass and sand substrate to feed (Itzkowitz, 1974b). *A. dussumieri* and *A. mata* are members of a highly specialized group that extract algae, diatoms and detritus from sand picked up in open, shallow areas. They have thick-walled, gizzard-like stomachs that break up the diatoms (Jones, 1968). The other species listed are schooling herbivores that cruise about close to the reef substrate, some occasionally in mixed-species schools, while feeding. Each species is usually found in a characteristic sub-zone of the reef, with some movement according to tides. Some acanthurids are among those species that form feeding schools to overcome the territorial aggression of other benthivorous fishes, as discussed in Chapter 2 (Sect.2.5.3).

Among the parrotfishes (Scaridae) some species form schools, especially when actively feeding. All scarids are diurnal benthivores, feeding mainly on algae (Randall, 1968; Hobson, 1974), but sometimes ingesting coral polyps along with the algae and fine calcareous material scraped from the surface of coral rocks with their strong beak-like dental plates (Hiatt and Strasburg, 1960). In the Caribbean Sea the feeding and social behaviour of *Scarus croicensis* have been studied intensively by several workers, including Winn and Bardach (1960), Itzkowitz (1974b) and Brawley and Adey (1977). Its social organization is variable. Some individuals defend territories in small groups, others are in small stationary groups, but are not territorial, and still others form mobile, feeding schools that may contain up to 500 fish (Ogden and Buckman, 1973). The latter feed on coral, and appear to avoid areas of grass (*Thalassia*) and open sand (Robertson et al., 1976). The feeding strategy of schooling *S. croicensis* near Puerto Rico may be typical. A group settles onto the coral substrate, each fish rasps and sucks rapidly and repeatedly at the bottom, and the group then rises and moves some distance in a cohesive school before settling for another bout of feeding. Barlow (1975) argued that this is an effective technique for exploiting food that is patchily distributed.

## 7.5 Evolution of Social Systems

Recent attempts to understand the evolution of vertebrate social systems have used the approach of relating the characteristics of a species' social organization to its ecological setting in the broadest sense. The assumption has been that its social structure is as much a product of a species' adaptive responses to its past and present environment as are its morphological and physiological characteristics. Attempts have been made to correlate social organization with a local population's

**Table 7.7.** Summary of common associations between fish social units and the basic requirements of all species

| Social unit | Resource exploitation | | Predator avoidance | Mating | Parental care |
|---|---|---|---|---|---|
| | Diet | Shelter | | | |
| *Single fish* a) Solitary | Mainly carnivorous, on or near substrate | None, or in substrate | Retreat to shelter; secretive or use spines, armour, large size, etc. | Temporary pairs or small groups | Usually none |
| b) Territorial mosaic | Variable, mainly benthos | In territory | Retreat to shelter | Temporary association in male's territory | Male only |
| *Male – female pair* | Benthos, omnivorous or coral polyps | Within home range | Retreat to shelter. Sharp spines | Little known, probably periodic | None |
| *Small group* | Benthos, omnivorous | Within territory | Retreat to shelter | Within group | By adult pair or male only |
| *School* | Variable, depending on habitat | Within home range or none | Individual distance decreases; variety of maneuvres | Within school, or temporary breeding territories | Usually none, male only in some |

habitat, diet, resting areas, and predators. Most of this work has concerned birds and mammals, with special emphasis recently on primates (e.g., Crook, 1965, 1970; Kummer, 1968; Struhsaker, 1969; Eisenberg et al., 1972; Geist, 1974; Jarman, 1974; Clutton-Brock, 1974; Clutton-Brock and Harvey, 1977). A similar approach with fishes is still rudimentary, although some promising beginnings have been made (Barlow, 1974a; Fricke, 1975b; Reese, 1975).

A species' most important requirements ("vital functions" in the terminology of Crook et al., 1976) can be grouped into four main categories: resource exploitation (including food and shelter), predator avoidance, mating, and rearing of young. If the last two are combined into reproduction, the resulting three requirements are the same as those listed in the Introduction to this book. Different sections of the book have dealt with the specific behavioural strategies used by fishes to meet these requirements. Here I consider the relationships between social systems in general and the four basic requirements. This is done in the form of two summarizing tables, where the four types of social unit discussed in this chapter (single fish, male–female pair, small group, and school) are first related to the basic requirements (Table 7.7), and then to the main behavioural mechanisms that operate within and between the social units (Table 7.8). Based on these two tables a number of tentative generalizations can be made. There are dangers of oversimplification in this approach. Summary tables cannot indicate the extensive variation in the social organization–ecology interplay that exists in some groups, for example the chaetodontids (Reese, 1975) and the acanthurids (Barlow, 1974a). Also, they are based on the specific examples discussed in earlier chapters; these are

**Table 7.8.** Summary of the common associations between fish social units and the behavioural mechanisms operating within and between the social units

| Social unit | Individual distance | Home range | Territory | Dominance |
|---|---|---|---|---|
| *Single fish*<br>a) Solitary | Large | Common, large. Some wander | Rare, only those with shelters | Does not apply |
| b) Territorial mosaic | Difficult to distinguish from territory | Same as territory | All-purpose | Dominant in territory, subordinate outside |
| *Male – female pair* | Small | Common, large. Some wander | Rare | None |
| *Small group* | Small | Similar to territory | Common, around sea anemone or other distinct benthic structure | Size-dependent hierarchy |
| *School* | Small, varies with behaviour | Usually large | None | None |

already a selective sampling of the available information. Yet I feel there is value in drawing up rough generalities of this sort, especially if the exercise leads to the formulation of questions that will focus the direction of future research.

The social units that appear to be most successful, in that they are found in virtually all aquatic habitats, are at opposite extremes in this series, that is, the solitary fish and the school. Solitary fishes are generally closely associated with the substrate, where they obtain their largely animal diet. They avoid predators by using substrate shelters or by their natural defences, such as large size, armour, and spines. Some roam widely over large home ranges; others appear to wander, although further evidence may show that even these species use familiar home ranges. Only those that stay close to their shelters, such as moray eels (Muraenidae), appear actively to defend an area around the shelter. In contrast, some schooling fishes live near the bottom, others in mid-water, still others near the surface. Their diet differs according to their size, feeding apparatus and habitat; some are substrate herbivores, some are piscivores, many are planktivores. They derive protection from attack by their fluid, changeable schooling behaviour. In general they are mobile, strong swimmers and move over extensive home ranges, although these latter features are more pronounced among strongly polarized schoolers than among species that are predominantly non-polarized.

The territorial mosaic system is also frequently seen, especially in shallow marine or freshwater areas where competition for limited resources on the substrate is stiff. Each individual stakes out a territory that it defends vigorously, especially against conspecifics, but often against members of other species as well. Generally all basic requirements are met within the individual's territory.

The long-term monogamous male–female pair and the small group (including usually one adult pair and some subadults) are more specialized social units, and are much less common than the other types. They are only well known so far from coral reef ecosystems. The small groups are closely, in some cases permanently, associated with conspicuous, somewhat isolated habitats on the substrate, such as giant sea anemones and small, discrete colonies of branching coral. All of their basic requirements are met in the immediate neighbourhood of the host structure. The monogamous pairs are so far only well documented in some chaetodontid species. They are also strongly substrate-bound, but have large home ranges.

Other recent proposals to correlate fish social structure with ecological factors are those of Barlow, Fricke, and Reese. Barlow (1974a) concluded that the variability in social systems among Central American cichlids and Pacific ocean acanthurids is basically a result of adaptations to different feeding strategies, but that predation and the physical environment (bottom topography, tidal changes, turbidity, etc.) have important short-term effects. Fricke (1975b) argued that the common types of social structure among Red Sea pomacentrids can be seen as behavioural adaptations to a number of ecological factors, the most important of which are: food, shelter, predators, abiotic factors, and the number of other conspecifics in the immediate area.

Finally, Reese (1975) described seven "behavioural ecological categories" among chaetodontids. These were distinguished on the basis of their social behaviour, space-related behaviour and diet. Interestingly, among the 20 species investigated, Reese found three (*C. auriga, C. lineolatus*, and *C. lunula*) that were so variable in their social relationships, that he was unable to classify them with confidence. They are seen alone, in pairs or in small groups at different locations, and Reese concluded that the ecological setting, that was different at each study site, was responsible for this variation. This seems to be an especially valuable discovery. With such a labile social structure, the emphasis in future research should perhaps be on a careful search for causal explanations of this link between ecology and social behaviour, especially in these variable species.

As in other areas of ethology, description must eventually be followed by causal and functional analysis. Detailed comparative descriptions have only been made of a few groups of species, and a vast amount remains to be done. The greatest value of broadly based comparisons across related species may be in the stimulus they provide for the formulation of precise hypotheses to be tested by future experimental work, both in the field and the laboratory.

# References

Abel EF (1961) Freiwasserstudien über das Fortpflanzungsverhalten des Mönchfisches *Chromis chromis* Linné, einem Vertreter der Pomacentriden im Mittelmeer. Z Tierpsychol 18:441-449

Abel EF (1964) Freiwasserstudien zur Fortpflanzungsethologie zweier Mittelmeerfische, *Blennius canevae* Vinc und *Blennius inaequalis* C V. Z Tierpsychol 21:205-222

Albrecht H (1968) Freiwasserbeobachtungen an Tilapien (Pisces, Cichlidae) in Ostafrika. Z Tierpsychol 25:377-394

Albrecht H (1969) Behaviour of four species of Atlantic damselfishes from Colombia, South America (*Abudefduf saxatilis, A taurus, Chromis multilineata, C cyanea*; Pisces, Pomacentridae). Z Tierpsychol 26:662-676

Alevizon WS (1976) Mixed schooling and its possible significance in a tropical Western Atlantic parrotfish and surgeonfish. Copeia 1976:796-798

Alexander R McN (1967) Functional design in fishes. Hutchinson, London, p 160

Allen GR (1975) The anemonefishes. Their classification and biology, 2nd edn. TFH Publ, Neptune City, NJ, p 352

Applegate VC (1950) Natural history of the sea lamprey, *Petromyzon marinus*, in Michigan. US Fish Wildl Serv Spec Sci Rep Fish 55:1-237

Arnold DC (1953) Observations on *Carapus acus* (Brunnich), (Jugulares, Carapidae). Pubbl Stn Zool Napoli 24:152-166

Arnold M, Kriesten K, Peters HM (1968) Die Haftorgane von *Tilapia*-Larven (Cichlidae, Teleostei). Z Zellforsch 91:248-260

Assem J van den (1967) Territory in the three-spined stickleback *Gasterosteus aculeatus* L. Behaviour Suppl 16:1-164

Atz JW (1958) A mouthful of babies. Anim Kingdom 61:182-186

Baerends GP, Baerends-van Roon JM (1950) An introduction to the study of the ethology of Cichlid fishes. Behaviour Suppl 1:1-243

Baerends GP, Brouwer R, Waterbolk HTJ (1955) Ethological studies on *Lebistes reticulatus* (Peters). I. An analysis of the male courtship pattern. Behaviour 8:249-334

Bahr K (1953) Beiträge zur Fortpflanzungsbiologie des Flußneunauges, *Petromyzon fluviatilis* L. Zool Jahrb Syst 82:58-69

Bailey RM (ed) (1970) A list of common and scientific names of fishes from the United States and Canada, 3rd edn. Am Fish Soc Spec Publ 6:1-150

Baldaccini NE (1973) An ethological study of reproductive behaviour including the colour patterns of the cichlid fish *Tilapia mariae* (Boulenger). Monit Zool Ital 7:247-290

Balon EK (1975) Reproductive guilds of fishes: a proposal and definition. J Fish Res Board Can 32:821-864

Banner A (1972) Use of sound in predation by young lemon sharks, *Negaprion brevirostris* (Poey). Bull Mar Sci 22:251-283

Bardach JE (1961) Transport of calcareous fragments by reef fishes. Science 133:98-99

Barlow GW (1961) Social behavior of the desert pupfish, *Cyprinodon macularius*, in the field and in the aquarium. Am Midl Nat 65:339-359

Barlow GW (1964) Ethology of the Asian teleost *Badis badis*. V. Dynamics of fanning and other parental activities, with comments on the behavior of the larvae and postlarvae. Z Tierpsychol 21:99-123

Barlow GW (1967) The functional significance of the split-head color pattern as exemplified in a leaf fish, *Polycentrus schomburgkii*. Ichthyologica 39:57-70

Barlow GW (1970) A test of appeasement and arousal hypotheses of courtship behavior in a cichlid fish, *Etroplus maculatus*. Z Tierpsychol 27:779-806

Barlow GW (1974a) Contrasts in social behavior between Central American cichlid fishes and coral-reef surgeon fishes. Am Zool 14:9-34

178

Barlow GW (1974b) Extraspecific imposition of social grouping among surgeonfishes (Pisces: Acanthuridae). J Zool Lond 174:333-340

Barlow GW (1975) On the sociobiology of four Puerto Rican parrotfishes (Scaridae). Mar Biol 33:281-293

Barlow GW (1976) The Midas cichlid in Nicaragua. In: Thorson TB (ed) Investigations of the ichthyofauna of Nicaraguan lakes. Univ Nebraska, Lincoln, pp 332-358

Barlow GW, Green RF (1970) The problems of appeasement and of sexual roles in the courtship behaviour of the blackchin mouthbreeder, *Tilapia melanotheron* (Pisces: Cichlidae). Behaviour 36:84-115

Barney RL, Anson BJ (1923) Life history and ecology of the orange-spotted sunfish, *Lepomis humilis*. Appendix XV, Rep US Comm Fish (1922) 1-16

Bastock M (1967) Courtship. A zoological study. Heinemann, London, p 220

Bauer J (1968) Vergleichende Untersuchungen zum Kontaktverhalten verschiedener Arten der Gattung *Tilapia* (Cichlidae, Pisces) und ihrer Bastarde. Z Tierpsychol 25:22-70

Baylis JR (1974) The behavior and ecology of *Herotilapia multispinosa* (Teleostei, Cichlidae). Z Tierpsychol 34:115-146

Baylis JR (1976a) A quantitative study of long-term courtship: I. Ethological isolation between sympatric populations of the midas cichlid, *Cichlasoma citrinellum*, and the arrow cichlid, *C zaliosum*. Behaviour 59:59-69

Baylis JR (1976b) A quantitative study of long-term courtship: II. A comparative study of the dynamics of courtship in two New World cichlid fishes. Behaviour 59:117-161

Belk MS (1975) Habitat partitioning in two tropical reef fishes, *Pomacentrus lividus* and *P albofasciatus*. Copeia 1975:603-607

Benson AA, Lee RF (1975) The role of wax in oceanic food chains. Sci Am 232(3):76-86

Benson AA, Muscatine L (1974) Wax in coral mucus: energy transfer from corals to reef fishes. Limnol Oceanogr 19:810-814

Bergmann H-H (1968) Eine deskriptive Verhaltensanalyse des Segelflossers (*Pterophyllum scalare* Cuv & Val, Cichlidae, Pisces). Z Tierpsychol 25:559-587

Bertelsen E (1951) The ceratioid fishes. Ontogeny, taxonomy, distribution and biology. Dana Rep 39:1-276

Beukema JJ (1963) Experiments on the effects of the hunger state and of a learning process on the risk of prey of the three-spined stickleback (*Gasterosteus aculeatus* L). Arch Néerl Zool 15:358-361

Black R, Wootton RJ (1970) Dispersion in a natural population of three-spined sticklebacks. Can J Zool 48:1133-1135

Bone Q (1975) Muscular and energetic aspects of fish swimming. In: Wu TYT, Brokaw CJ, Brennan C (eds) Swimming and flying in nature, vol II. Plenum, New York, pp 493-528

Bowen ES (1931) The role of the sense organs in aggregations of *Ameiurus melas*. Ecol Monogr 1:1-35

Braddock JC, Braddock ZI (1959) The development of nesting behaviour in the Siamese fighting fish *Betta splendens*. Anim Behav 7:222-232

Brawley SH, Adey WH (1977) Territorial behavior of threespot damselfish (*Eupomacentrus planifrons*) increases reef algal biomass and productivity. Environ Biol Fish 2:45-51

Brawn VM (1961) Reproductive behaviour of the cod (*Gadus callarias* L). Behaviour 18:177-198

Breder CM (1926) The locomotion of fishes. Zoologica 4:159-297

Breder CM (1935) The reproductive habits of the common catfish, *Ameiurus nebulosus* (Le Sueur), with a discussion of their significance in ontogeny and phylogeny. Zoologica 19:143-185

Breder CM (1936) The reproductive habits of the North American sunfishes (Family Centrarchidae). Zoologica 21:1-48

Breder CM (1939) Variations in the nesting habits of *Ameiurus nebulosus* (Le Sueur). Zoologica 24:367-378

Breder CM (1941a) On the reproduction of *Opsanus beta* Goode and Bean. Zoologica 26:229-232

Breder CM (1941b) On the reproductive behavior of the sponge blenny, *Paraclinus marmoratus* (Steindachner). Zoologica 26:233-236

Breder CM (1946) An analysis of the deceptive resemblances of fishes to plant parts, with critical remarks on protective coloration, mimicry and adaptation. Bull Bingham Oceanogr Collect 10(2):1-49

Breder CM (1949) On the behavior of young *Lobotes surinamensis*. Copeia 1949:237-242

179

Breder CM (1951) Studies on the structure of the fish school. Bull Am Mus Nat Hist 98:1-27

Breder CM (1959) Studies on social groupings in fishes. Bull Am Mus Hist 117:393-482

Breder CM (1963) Defensive behavior and venom in *Scorpaena* and *Dactylopterus*. Copeia 1963:698-700

Breder CM (1965) Vortices and fish schools. Zoologica 50:97-114

Breder CM (1967) On the survival value of fish schools. Zoologica 52:25-40

Breder CM (1976) Fish schools as operational structures. Fish Bull 74:471-502

Breder CM, Coates CW (1933) Reproduction and eggs of *Pomacentrus leucoris* Gilbert. Am Mus Novit 612:1-6

Breder CM, Rosen DE (1966) Modes of reproduction in fishes. Natural History Press, Garden City, NY, p 941

Brestowsky M (1968) Vergleichende Untersuchungen zur Elternbindung von *Tilapia*-Jungfischen (Cichlidae, Pisces). Z Tierpsychol 25:761-828

Brichard P (1975) Réflexions sur le choix de la nidification ou de l'incubation buccale comme mode de reproduction chez certaines populations de poissons Cichlides du Lac Tanganyika. Rev Zool Afr 89:871-888

Briggs JC (1953) The behavior and reproduction of salmonid fishes in a small coastal stream. Calif Dep Fish Game Fish Bull 94:1-62

Brightwell LR (1953) Further notes on the hermit crab *Eupagurus bernhardus* and associated animals. Proc Zool Soc Lond 123:61-64

Brock VE, Chamberlain TC (1968) A geological and ecological reconnaissance off western Oahu, Hawaii, principally by means of the research submarine "Asherah". Pac Sci 22:373-394

Brock VE, Riffenburgh RH (1960) Fish schooling: a possible factor in reducing predation. J Cons 25:307-317

Brockmann HJ (1973) The function of poster-coloration in the Beaugregory, *Eupomacentrus leucostictus* (Pomacentridae, Pisces). Z Tierpsychol 33:13-34

Brooks JL (1968) The effects of prey size selection by lake planktivores. Syst Zool 17:273-291

Brown JH, Cantrell MA, Evans SM (1973) Observations on the behaviour and coloration of some coral reef fish (Family: Pomacentridae). Mar Behav Physiol 2:63-71

Brown JL (1975) The evolution of behavior. Norton, New York, p 761

Budker P (1971) The life of sharks. Columbia Univ, New York, p 222

Burchard J (1967) The family Cichlidae. In: Reed W (ed) Fish and fisheries of northern Nigeria. Min Agric No Nigeria, Zaria, pp 123-143

Burt WH (1943) Territoriality and home range concepts as applied to mammals. J Mammal 24:346-352

Campbell AC (1976) The coral seas. Orbis, London, p 128

Carranza J, Winn HE (1954) Reproductive behavior of the blackstripe topminnow, *Fundulus notatus*. Copeia 1954:273-278

Case B (1970) Spawning behaviour of the chestnut lamprey (*Ichthyomyzon castaneus*). J Fish Res Board Can 27:1872-1874

Chadwick HC (1929) Feeding habits of the angler-fish, *Lophius piscatorius*. Nature (London) 124:337

Chien AK, Salmon M (1972) Reproductive behavior of the angelfish, *Pterophyllum scalare*. I. A quantitative analysis of spawning and parental behavior. Forma Functio 5:45-74

Choat JH, Robertson DR (1975) Protogynous hermaphroditism in fishes of the family Scaridae. In: Reinboth R (ed) Intersexuality in the animal kingdom. Springer, Berlin Heidelberg New York, pp 263-283

Cichocki F (1977) Tidal cycling and parental behavior of the cichlid fish, *Biotodoma cupido*. Environ Biol Fish 1:159-169

Clark CF (1950) Observations on the spawning habits of the northern pike, *Esox lucius*, in northwestern Ohio. Copeia 1950:285-288

Clark E, Aronson LR (1951) Sexual behavior in the guppy, *Lebistes reticulatus* (Peters). Zoologica 36:49-66

Clark E, Aronson LR, Gordon M (1954) Mating behavior patterns in two sympatric species of Xiphophorin fishes: their inheritance and significance in sexual isolation. Bull Am Mus Nat Hist 103:135-226

Clark FN (1938) Grunion in southern California. Cal Fish Game 24:49-54

Clark FW, Keenleyside MHA (1967) Reproductive isolation between the sunfish *Lepomis gibbosus* and *L macrochirus*. J Fish Res Board Can 24:495-514

Clarke TA (1970) Territorial behavior and population dynamics of a pomacentric fish, the garibaldi, *Hypsypops rubicunda*. Ecol Monogr 40:189-212

Clarke TA (1971) Territory boundaries, courtship, and social behavior in the garibaldi, *Hypsypops rubicunda* (Pomacentridae). Copeia 1971:295-299

Clarke TA, Flechsig AO, Grigg RW (1967) Ecological studies during project Sealab II. Science 157:1381-1389

Clarke WD (1963) Function of bioluminescence in mesopelagic organisms. Nature (London) 198:1244-1246

Clutton-Brock TH (1974) Primate social organisation and ecology. Nature (London) 250:539-542

Clutton-Brock TH, Harvey PH (1977) Primate ecology and social organization. J Zool Lond 183:1-39

Coeckelberghs V (1975) Territorial, spawning and parental behaviour of *Lamprologus brichardi* Poll 1974 (Pisces, Cichlidae). Ann Soc R Zool Belg 105:73-86

Cole JE, Ward JA (1969) The communicative function of pelvic fin-flickering in *Etroplus maculatus* (Pisces, Cichlidae). Behaviour 35:179-199

Cole JE, Ward JA (1970) An analysis of parental recognition by the young of the cichlid fish, *Etroplus maculatus* (Bloch). Z Tierpsychol 27:156-176

Coles RJ (1915) Notes on the sharks and rays of Cape Lookout, N C. Proc Biol Soc Wash 28:89-94

Colgan P (1973) Motivational analysis of fish feeding. Behaviour 45:38-66

Colgan P (1974) Burying experiments with the banded killifish, *Fundulus diaphanus*. Copeia 1974:258-259

Colgan P, Ealey D (1973) Role of woody debris in nest site selection by pumpkinseed sunfish, *Lepomis gibbosus*. J Fish Res Board Can 30:853-856

Collette BB, Talbot FH (1972) Activity patterns of coral reef fishes with emphasis on nocturnal-diurnal changeover. Nat Hist Mus Los Angeles Cty Sci Bull 14:98-124

Collins HL (1972) Mouth brooding behavior in the substrate spawning cichlid, *Tilapia sparrmani*. J Minn Acad Sci 38:17-18

Constantz GD (1975) Behavioral ecology of mating in the male Gila topminnow, *Poeciliopsis occidentalis* (Cyprinodontiformes; Poeciliidae). Ecology 56:966-973

Cott HB (1940) Adaptive coloration in animals. Methuen, London, p 508

Cousteau J-Y, Cousteau P (1970) The shark: splendid savage of the sea. Doubleday, New York, p 277

Crook JH (1965) The adaptive significance of avian social organizations. Symp Zool Soc Lond 14:181-218

Crook JH (1970) Social organization and the environment: aspects of contemporary social ethology. Anim Behav 18:197-209

Crook JH, Ellis JE, Goss-Custard JD (1976) Mammalian social systems: structure and function. Anim Behav 24:261-274

Cross DG (1969) Aquatic weed control using grass carp. J Fish Biol 1:27-30

Crossman EJ (1959) A predator-prey interaction in freshwater fish. J Fish Res Board Can 16:269-281

Crossman EJ (1977) Displacement, and home range movements of muskellunge determined by ultrasonic tracking. Environ Biol Fish 1:145-158

Curio E (1976) The ethology of predation. Zoophys Ecol, vol 7. Springer, Berlin Heidelberg New York, p 250

Daugherty CH, Daugherty LB, Blair AP (1976) Visual and olfactory stimuli in the feeding behavior of darters (*Etheostoma*) inhabiting clear and muddy water. Copeia 1976: 380-382

Dawson JA (1963) The oral cavity, the "jaws" and the horny teeth of *Myxine glutinosa*. In: Brodal A, Fange R (eds) The biology of *Myxine*. Universitetsforlaget, Oslo, pp 231-255

De Bont AF (1949) La reproduction en étangs des *Tilapia melanopleura* (Dum) et *macrochir* (Blgr). C R Conf Pisc Anglo-Belge, Elisabethville, pp 303-312

Deelder CL (1951) A contribution to the knowledge of the stunted growth of perch (*Perca fluviatilis* L) in Holland. Hydrobiologia 3:357-378

Dendy JS, Scott DC (1953) Distribution, life history, and morphological variations of the southern brook lamprey, *Ichthyomyzon gagei*. Copeia 1953:152-162

Denison RH (1961) Feeding mechanisms of Agnatha and early Gnathostomes. Am Zool 1:177-181

De Silva SS, Kortmulder K, Wijeyaratne MJS (1977) A comparative study of the food and feeding habits of *Puntius bimaculatus* and *P titteya* (Pisces, Cyprinidae). Neth J Zool 27:253-263

Diana JS, MacKay WC, Ehrman M (1977) Movements and habitat preference of northern pike (*Esox lucius*) in Lac Ste Anne, Alberta. Trans Am Fish Soc 106:560-565

Dijkgraaf S, Kalmijn AJ (1963) Untersuchungen über die Funktion der Lorenzinischen Ampullen an Haifischen. Z Vergl Physiol 47:438-456

Dill LM (1977) Refraction and the spitting behavior of the archerfish (*Toxotes chatareus*). Behav Ecol Sociobiol 2:169-184

Doan KH (1938) Observations on dogfish (*Amia calva*) and their young. Copeia 1938:204

Ebersole JP (1977) The adaptive significance of interspecific territoriality in the reef fish *Eupomacentrus leucostictus*. Ecology 58:914-920

Echelle AA (1973) Behavior of the pupfish, *Cyprinodon rubrofluviatilis*. Copeia 1973:68-76

Edmunds M (1974) Defence in animals: a survey of anti-predator defences. Longman, Harlow, p 357

Eggers DM (1976) Theoretical effect of schooling by planktivorous fish predators on rate of prey consumption. J Fish Res Board Can 33:1964-1971

Ehrlich PR, Ehrlich AH (1973) Coevolution: heterotypic schooling in Caribbean reef fishes. Am Nat 107:157-160

Ehrlich PR, Talbot FH, Russell BC, Anderson GRV (1977) The behaviour of chaetodontid fishes with special reference to Lorenz's "poster colouration" hypothesis. J Zool Lond 183:213-228

Eibl-Eibesfeldt I (1955) Über Symbiosen, Parasitismus und andere zwischenartliche Beziehungen bei tropischen Meeresfischen. Z Tierpsychol 12:203-219

Eibl-Eibesfeldt I (1959) Der Fisch *Aspidontus taeniatus* als Nachahmer des Putzers *Labroides dimidiatus*. Z Tierpsychol 16:19-25

Eibl-Eibesfeldt I (1961) Eine Symbiose zwischen Fischen (*Siphamia versicolor*) und Seeigeln. Z Tierpsychol 18:56-59

Eibl-Eibesfeldt I (1962) Freiwasserbeobachtungen zur Deutung des Schwarmverhaltens verschiedener Fische. Z Tierpsychol 19:165-182

Eibl-Eibesfeldt I (1975) Ethology, the biology of behavior, 2nd edn. Holt, Rinehart and Winston, New York, p 625

Eibl-Eibesfeldt I, Hass H (1959) Erfahrungen mit Haien. Z Tierpsychol 16:733-746

Eigenmann CH (1912) The freshwater fishes of British Guiana, including a study of the ecological grouping of species and the relation of the fauna of the plateau to that of the lowlands. Mem Carnegie Mus 5:1-578

Eisenberg JF, Muckenhirn NA, Rudran R (1972) The relation between ecology and social structure in primates. Science 176:863-874

Elson PF (1957) The importance of size in the change from parr to smolt in Atlantic salmon. Can Fish Cult 21:1-6

Emlen JM (1973) Ecology: an evolutionary approach. Addison-Wesley, Don Mills, p 493

Fabricius E, Gustafson K-J (1954) Further aquarium observations on the spawning behaviour of the char, *Salmo alpinus* L. Rep Inst Freshwater Res Drottningholm 35:58-104

Fabricius E, Gustafson K-J (1958) Some new observations on the spawning behaviour of the pike, *Esox lucius* L. Rep Inst Freshwater Res Drottningholm 39:23-54

Farmer GJ, Beamish FWH (1973) Sea lamprey (*Petromyzon marinus*) predation on freshwater teleosts. J Fish Res Board Can 30:601-605

Fernald RD, Hirata NR (1977) Field study of *Haplochromis burtoni*: quantitative behavioural observations. Anim Behav 25:964-975

Fiedler K (1964) Verhaltensstudien an Lippfischen der Gattung *Crenilabrus* (Labridae, Perciformes). Z Tierpsychol 21:521-591

Fish GR (1955) The food of *Tilapia* in East Africa. Uganda J 19:85-89

Fishelson L (1964) Observations on the biology and behaviour of Red Sea coral fishes. Bull Isr Sea Fish Res Stn Isr 37:11-26

Fishelson L (1966) Cichlidae of the genus *Tilapia* in Israel. Bull Fish Cult Isr 18:67-80

Fishelson L (1970) Behaviour and ecology of a population of *Abudefduf saxatilis* (Pomacentridae, Teleostei) at Eilat (Red Sea). Anim Behav 18:225-237

Fishelson L (1977) Sociobiology of feeding behavior of coral fish along the coral reef of the Gulf of Elat (= Gulf of Aqaba), Red Sea. Isr J Zool 26:114-134

182

Fishelson L, Popper D, Avidor A (1974) Biosociology and ecology of pomacentrid fishes around the Sinai peninsula (northern Red Sea). J Fish Biol 6:119-133

FitzGerald GJ, Keenleyside MHA (1978) The effects of numerical density of adult fish on reproduction and parental behavior in the convict cichlid fish *Cichlasoma nigrofasciatum* (Günther). Can J Zool 56:1367-1371

Foerster RE (1968) The sockeye salmon, *Oncorhynchus nerka*. Bull Fish Res Board Can 162:1-422

Forselius S (1957) Studies of anabantid fishes. I. A qualitative description of the reproductive behaviour in territorial species investigated under laboratory conditions with special regard to genus *Colisa*. Zool Bid Uppsala 32:93-302

Foster JR (1977) The role of breeding behavior and habitat preferences on the reproductive isolation of three allopatric populations of ninespine stickleback, *Pungitius pungitius*. Can J Zool 55:1601-1611

Foster NR (1967) Trends in the evolution of reproductive behavior in killifishes. Stud Trop Oceanogr 5:549-566

Foster NR (1973) Behavior, development, and early life history of the Asian needlefish, *Xenentodon cancila*. Proc Acad Nat Sci Philadelphia 125:77-88

Fowler HW (1923) Spawning habits of sunfishes, basses etc. Fish Cult 2:226-228

Fricke HW (1970) Ein mimetisches Kollektiv — Beobachtungen an Fischschwärmen, die Seeigel nachahmen. Mar Biol 5:307-314

Fricke HW (1971) Fische als Feinde tropischer Seeigel. Mar Biol 9:328-338

Fricke HW (1973a) Behavior as part of ecological adaptation. Helgol Wiss Meeresunters 24:120-144

Fricke HW (1973b) Eine Fische-Seeigel-Partnerschaft. Untersuchungen optischer Reizparameter beim Formenerkennen. Mar Biol 19:290-297

Fricke HW (1973c) Ökologie und Sozialverhalten des Korallenbarsches *Dascyllus trimaculatus* (Pisces, Pomacentridae). Z Tierpsychol 32:225-256

Fricke HW (1974) Öko-Ethologie des monogamen Anemonenfisches *Amphiprion bicinctus* (Freiwasseruntersuchung aus dem Roten Meer). Z Tierpsychol 36:429-512

Fricke HW (1975a) Lösen einfacher Probleme bei einem Fisch (Freiwasserversuche an *Balistes fuscus*). Z Tierpsychol 38:18-33

Fricke HW (1975b) Sozialstruktur und ökologische Spezialisierung von verwandten Fischen (Pomacentridae). Z Tierpsychol 39:492-520

Fricke HW (1977) Aggression: Motor zum Überleben. Geo 1977(8):42-64

Fricke H, Fricke S (1977) Monogamy and sex change by aggressive dominance in coral reef fish. Nature (London) 266:830-832

Fricke HW, Holzberg S (1974) Social units and hermaphroditism in a pomacentrid fish. Naturwissenschaften 61:367-368

Frisch von K (1938) Zur Psychologie des Fischschwarmes. Naturwissenschaften 26:601-606

Frisch von K (1942) Über einen Schreckstoff der Fischhaut und seine biologische Bedeutung. Z Vergl Physiol 29:46-145

Friswold C (1944) *Loricaria parva*, further observations. Aquarium 13:67-70

Fryer G (1959) The trophic interrelationships and ecology of some littoral communities of Lake Nyasa with special reference to the fishes, and a discussion of the evolution of a group of rock-frequenting Cichlidae. Proc Zool Soc Lond 132:153-281

Fryer G (1969) Speciation and adaptive radiation in African lakes. Verh Int Ver Limnol 17:303-322

Fryer G (1977) Evolution of species flocks of cichlid fishes in African lakes. Z Zool Syst Evolutionsforsch 15:141-165

Fryer G, Iles TD (1972) The cichlid fishes of the Great Lakes of Africa. TFH Public, Neptune City, New Jersey, p 641

Galbraith MG (1967) Size-selective predation on *Daphnia* by rainbow trout and yellow perch. Trans Am Fish Soc 96:1-10

Gale WF, Gale CA (1976) Selection of artificial spawning sites by the spotfin shiner (*Notropis spilopterus*). J Fish Res Board Can 33:1906-1913

Gale WF, Gale CA (1977) Spawning habits of spotfin shiner (*Notropis spilopterus*) — a fractional, crevice spawner. Trans Am Fish Soc 106:170-177

Gandolfi G, Mainardi D, Rossi AC (1968) La reazione di paura e lo svantaggio individuale dei pesci allarmisti (esperimenti con modelli). Ist Lomb Accad Sci Lett Rend Sci B 102:8-14

Geist V (1974) On the relationship of social evolution and ecology of ungulates. Am Zool 14:205-220

Gerald JW (1971) Sound production during courtship in six species of sunfish (Centrarchidae). Evolution 25:75-87

Gibson RN (1969) The biology and behaviour of littoral fish. Oceanogr Mar Biol Ann Rev 7:367-410

Gilbert PW (1963) Sharks and survival. Heath, Boston, p 578

Gosline WA (1971) Functional morphology and classification of teleostean fishes. Univ Press of Hawaii, Honolulu, p 208

Gosse J-P (1975) Révision du genre *Geophagus*. Acad R Sci Outre Mer Cl Sci Nat Med XIX:1-172

Gotto RV (1969) Marine animals; partnerships and other associations. Elsevier, New York, p 96

Graefe G (1964) Zur Anemonen-Fisch-Symbiose, nach Freilanduntersuchungen bei Eilat/Rotes Meer. Z Tierpsychol 21:468-485

Gray J (1953) The locomotion of fishes. In: Essays in marine biology. R Elmhirst Mem Lect. Oliver and Boyd, Edinburgh, pp 1-16

Gray J (1968) Animal locomotion. Weidenfeld and Nicolson, London, p 479

Greenberg B, Zijlstra JJ, Baerends GP (1965) A quantitative description of the behaviour changes during the reproductive cycle of the cichlid fish *Aequidens portalegrensis* Hensel. Proc K Ned Akad Wet 68:135-149

Greenwood PH (1974) The cichlid fishes of Lake Victoria, East Africa: the biology and evolution of a species flock. Bull Br Mus Nat Hist (Zool) Suppl 6:1-134

Greenwood PH, Thomson KS (1960) The pectoral anatomy of *Pantodon buchholzi* Peters (a freshwater flying fish) and the related Osteoglossidae. Proc Zool Soc Lond 135:283-301

Griswold BL, Smith LL (1972) Early survival and growth of the ninespine stickleback, *Pungitius pungitius*. Trans Am Fish Soc 101:350-352

Gudger EW (1916) The gaff-topsail (*Felichthys felis*), a sea catfish that carries its eggs in its mouth. Zoologica 2:125-158

Gudger EW (1930a) On the alleged penetration of the human urethra by an Amazonian catfish called candiru, with a review of the allied habits of other members of the family Pygidiidae. Am J Surg 8:170-188, 443-457

Gudger EW (1930b) The candirú, the only vertebrate parasite of man. Hoeber, New York

Hagelin L-O (1959) Further aquarium observations on the spawning habits of the river lamprey (*Petromyzon fluviatilis*). Oikos 10:50-64

Hagelin L-O, Steffner N (1958) Notes on the spawning habits of the river lamprey (*Petromyzon fluviatilis*). Oikos 9:221-238

Hagen DW (1967) Isolating mechanisms in threespine sticklebacks (*Gasterosteus*). J Fish Res Board Can 24:1637-1692

Hall DD (1968) A qualitative analysis of courtship and reproductive behavior in the paradise fish, *Macropodus opercularis* (Linnaeus). Z Tierpsychol 25:834-842

Hall DD, Miller RJ (1968) A qualitative study of courtship and reproductive behavior in the pearl gourami, *Trichogaster leeri* (Bleeker). Behaviour 32:70-84

Hamilton WD (1971) Geometry for the selfish herd. J Theor Biol 31:295-311

Hanamura N (1966) Salmon of the north Pacific Ocean. Part III. A review of the life history of north Pacific salmon. I. Sockeye salmon in the Far East. Bull Int North Pac Fish Comm 18:1-27

Hanson AJ, Smith HD (1967) Mate selection in a population of sockeye salmon (*Oncorhynchus nerka*) of mixed age-groups. J Fish Res Board Can 24:1955-1977

Hara TJ (1971) Chemoreception. In: Hoar WS, Randall DJ (eds) Fish physiology, vol 5. Academic Press, New York, pp 79-120

Hardisty MW, Potter IC (1971) The general biology of adult lampreys. In: Hardisty MW, Potter IC (eds) The biology of lampreys, vol 1. Academic Press, New York London, pp 127-206

Harris JE (1936) The role of the fins in the equilibrium of the swimming fish. I. Wind-tunnel tests on a model of *Mustelus canis* (Mitchill). J Exp Biol 13:476-493

Harris JE (1938) The role of the fins in the equilibrium of the swimming fish. II. The role of the pelvic fins. J Exp Biol 15:32-47

Harris JE (1953) Fin patterns and mode of life in fishes. In: Essays in marine biology. R Elmhirt Mem Lect. Oliver and Boyd, Edinburgh, pp 17-28

Harris VA (1960) On the locomotion of the mud-skipper *Periophthalmus koelreuteri* (Pallas) (Gobiidae). Proc Zool Soc Lond 134:107-135

Hart JL (1973) Pacific fishes of Canada. Bull Fish Res Board Can 180:1-740

Hart JL, Tester AL (1934) Quantitative studies on herring spawning. Trans Am Fish Soc 64:307-312

184

Hartline AC, Hartline PH, Szmant AM, Flechsig AO (1972) Escape response in a pomacentrid reef fish, *Chromis cyaneus*. Sci Bull Los Angeles Cty Mus 14:93-97

Hartman GF (1963) Observations on behaviour of juvenile brown trout in a stream aquarium during winter and spring. J Fish Res Board Can 20:769-787

Hartman GF (1969) Reproductive biology of the Gerrard stock rainbow trout. In: Northcote TG (ed) Symposium on salmon and trout in streams. MacMillan Lectures in Fisheries, Univ B C Vancouver, pp 53-67

Hartman GF (1970) Nest digging behavior of rainbow trout (*Salmo gairdneri*). Can J Zool 48:1458-1462

Hastings JW (1971) Light to hide by: ventral luminescence to camouflage the silhouette. Science 173:1016-1017

Hawkins JM (1974) Habitat preference in the central mudminnow, *Umbra limi*. M Sc Thesis, Univ Western Ontario, London, p 67

Heard WR (1966) Observations on lampreys in the Naknek River system of southwest Alaska. Copeia 1966:332-339

Heard WR (1972) Spawning behavior of pink salmon on an artificial redd. Trans Am Fish Soc 101:276-283

Hediger H (1955) Studies of the psychology and behaviour of captive animals in zoos and circuses. Butterworth, London, p 166

Herald ES (1959) From pipefish to seahorse — a study of phylogenetic relationships. Proc Calif Acad Sci Ser 4 29:465-473

Herting GE, Witt A (1967) The role of physical fitness of forage fishes in relation to their vulnerability to predation by bowfin (*Amia calva*). Trans Am Fish Soc 96:427-430

Heymer A, Auger de Ferret C (1976) Zur Ethologie des Mittelmeer-Schleimfisches *Blennius rouxi* Cocco 1833. Z Tierpsychol 41:121-141

Hiatt RW, Brock VE (1948) On the herding of prey and the schooling of the black skipjack, *Euthynnus yaito* Kishinouye. Pac Sci 2:297-298

Hiatt RW, Strasburg DW (1960) Ecological relationships of the fish fauna on coral reefs of the Marshall Islands. Ecol Monogr 30:65-127

Hickling CF (1961) Tropical inland fisheries. Longmans, London, p 287

Hildemann WH (1959) A cichlid fish, *Symphysodon discus*, with unique nurture habits. Am Nat 93:27-34

Hinton S (1962) Unusual defense movements in *Scorpaena plumieri mystes*. Copeia 1962:842

Hoar WS (1958) The evolution of migratory behaviour among juvenile salmon of the genus *Oncorhynchus*. J Fish Res Board Can 15:391-428

Hoar WS (1976) Smolt transformation: evolution, behavior, and physiology. J Fish Res Board Can 33:1233-1252

Hobson ES (1963) Feeding behavior in three species of sharks. Pac Sci 17:171-194

Hobson ES (1966) Visual orientation and feeding in seals and sea lions. Nature (London) 210:326-327

Hobson ES (1968a) Predatory behavior of some shore fishes in the Gulf of California. Res Rep US Fish Wildl Serv 73:1-92

Hobson ES (1968b) Coloration and activity of fishes, day and night. Underwater Nat 5(3):6-11

Hobson ES (1972) Activity of Hawaiian reef fishes during the evening and morning transitions between daylight and darkness. Fish Bull 70:715-740

Hobson ES (1974) Feeding relationships of teleostean fishes on coral reefs in Kona, Hawaii. Fish Bull 72:915-1031

Hobson ES, Chess JR (1978) Trophic relationships among fishes and plankton in the lagoon at Enewetak Atoll, Marshall Islands. Fish Bull 76:133-153

Holzberg S (1973) Beobachtungen zur Ökologie und zum Sozialverhalten des Korallenbarsches *Dascyllus marginatus* Rüppell (Pisces: Pomacentridae). Z Tierpsychol 33:492-513

Hoogland RD (1951) On the fixing-mechanism in the spines of *Gasterosteus aculeatus* L. Proc K Ned Akad Wet 54:171-180

Hoogland R, Morris D, Tinbergen N (1956) The spines of sticklebacks (*Gasterosteus* and *Pygosteus*) as means of defence against predators (*Perca* and *Esox*). Behaviour 10:205-236

Hourston AS, Rosenthal H (1976) Sperm density during active spawning of Pacific herring (*Clupea harengus pallasi*). J Fish Res Board Can 33:1788-1790

Hourston AS, Rosenthal H, Stacey N (1977) Observations on spawning behaviour of Pacific herring in captivity. Meeresforschung 25:156-162

Hoyt JW (1975) Hydrodynamic drag reduction due to fish slimes. In: Wu TYT, Brokaw CJ, Brennan C (eds) Swimming and flying in nature, vol 2. Plenum, New York, pp 653-672

Hubbs CL (1933) Observations on the flight of fishes, with a statistical study of the flight of the Cypselurinae and remarks on the evolution of the flight of fishes. Mich Acad Sci Arts Lett 17:575-611

Hubbs CL (1937) Further observations and statistics on the flight of fishes. Mich Acad Sci Arts Lett 22:641-660

Humphries DA, Driver PM (1970) Protean defence by prey animals. Oecologia 5:285-302

Hurley AC, Hartline PH (1974) Escape response in the damselfish *Chromis cyanea* (Pisces: Pomacentridae): a quantitative study. Anim Behav 22:430-437

Hynes HBN (1970) The ecology of running waters. Univ Toronto Press, Toronto, p 555

Idyll CP (1971) Abyss, the deep sea and the creatures that live in it. Rev ed. Crowell, New York, p 428

Iersel JJA van (1953) An analysis of the parental behaviour of the male three-spined stickleback (*Gasterosteus aculeatus* L). Behaviour Suppl III:1-159

Iersel JJA van (1958) Some aspects of territorial behaviour of the male three-spined stickleback. Arch Néerl Zool 13, Suppl 1:383-400

Ingram WM, Odum E (1941) Nests and behavior of *Lepomis gibbosus* (Linnaeus) in Lincoln Pond, Rensselaerville, New York. Am Mdl Nat 26:182-193

Itzkowitz M (1969) Observations on the breeding behavior of *Cyprinodon elegans* in swift water. Tex J Sci 21:229-231

Itzkowitz M (1974a) The effects of other fish on the reproductive behavior of the male *Cyprinodon variegatus* (Pisces: Cyprinodontidae). Behaviour 48:1-22

Itzkowitz M (1974b) A behavioural reconnaissance of some Jamaican reef fishes. Zool J Linn Soc 55:87-118

Janssen J (1976) Feeding modes and prey size selection in the alewife (*Alosa pseudoharengus*). J Fish Res Board Can 33:1972-1975

Janssen J (1978) Feeding-behavior repertoire of the alewife, *Alosa pseudoharengus*, and the ciscoes *Coregonus hoyi* and *C artedii*. J Fish Res Board Can 35:249-253

Jarman PJ (1974) The social organisation of antelope in relation to their ecology. Behaviour 48:215-267

Jenkins TM (1969) Social structure, position choice and microdistribution of two trout species (*Salmo trutta* and *Salmo gairdneri*) resident in mountain streams. Anim Behav Monogr 2:57-123

Jenni DA (1972) Effects of conspecifics and vegetation on nest site selection in *Gasterosteus aculeatus* L. Behaviour 42:97-118

Jones JW (1959) The salmon. Collins, London, p 192

Jones JW, King GM (1950) Further experimental observations on the spawning of the Atlantic salmon (*Salmo salar* Linn). Proc Zool Soc Lond 120:317-323

Jones JW, King GM (1952) The spawning of the male salmon parr (*Salmo salar* Linn juv). Proc Zool Soc Lond 122:615-619

Jones RS (1968) Ecological relationships in Hawaiian and Johnston Island Acanthuridae (surgeonfishes). Micronesica 4:309-361

Kalleberg H (1958) Observations in a stream tank of territoriality and competition in juvenile salmon and trout (*Salmo salar* L and *S trutta* L). Rep Inst Freshwater Res Drottningholm 39:55-98

Kalmijn AJ (1971) The electric sense of sharks and rays. J Exp Biol 55:371-383

Kalmijn AJ (1974) The detection of electric fields from inanimate and animate sources other than electric organs. In: Fessard A (ed) Handbook of sensory physiology, vol III. Springer, Berlin Heidelberg New York, pp 147-200

Karplus I, Szlep R, Tsurnamal M (1972) Associative behavior of the fish *Cryptocentrus cryptocentrus* (Gobiidae) and the pistol shrimp *Alpheus djiboutensis* (Alpheidae) in artificial burrows. Mar Biol 15:95-104

Kaufman L (1976) Feeding behavior and functional coloration of the Atlantic trumpetfish, *Aulostomus maculatus*. Copeia 1976:377-378

Keast A, Webb D (1966) Mouth and body form relative to feeding ecology in the fish fauna of a small lake, Lake Opinicon, Ontario. J Fish Res Board Can 23:1845-1874

Keenleyside MHA (1955) Some aspects of the schooling behaviour of fishes. Behaviour 8:183-248

186

Keenleyside MHA (1962) Skin-diving observations of Atlantic salmon and brook trout in the Miramichi River, New Brunswick. J Fish Res Board Can 19:625-634

Keenleyside MHA (1967) Behavior of male sunfishes (genus *Lepomis*) towards females of three species. Evolution 21:688-695

Keenleyside MHA (1972a) The behaviour of *Abudefduf zonatus* (Pisces, Pomacentridae) at Heron Island, Great Barrier Reef. Anim Behav 20:763-774

Keenleyside MHA (1972b) Intraspecific intrusions into nests of spawning longear sunfish (Pisces: Centrarchidae). Copeia 1972:272-278

Keenleyside MHA (1978a) Reproductive isolation between pumpkinseed (*Lepomis gibbosus*) and longear sunfish (*L megalotis*) (Centrarchidae) in the Thames River, southwestern Ontario. Can J Zool 35:131-135

Keenleyside MHA (1978b) Parental care behavior in fishes and birds. In: Reese ES, Lighter F (eds) Contrasts in behavior. Wiley, New York, pp 1-19

Keenleyside MHA, Prince CE (1976) Spawning-site selection in relation to parental care of eggs in *Aequidens paraguayensis* (Pisces: Cichlidae). Can J Zool 54:2135-2139

Kelly HA (1924) *Amia calva* guarding its young. Copeia 1924:73-74

Kelley WE, Atz JW (1964) A pygidiid catfish that can suck blood from goldfish. Copeia 1964:702-704

Klausewitz W (1964) Der Lokomotionsmodus der Flügelrochen (Myliobatoidei). Zool Anz 173:111-120

Kleerekoper H (1969) Olfaction in fishes. Indiana Univ Press, Bloomington, p 222

Kleerekoper H, Mogensen J (1963) Role of olfaction in the orientation of *Petromyzon marinus*. I. Response to a single amine in prey's body odor. Physiol Zool 36:347-360

Knipper H (1953) Beobachtungen an jungen *Plotosus anguillaris* (Bloch) (Pisces Nematognathi Plotosidae). Veroeff Ueberseemus Bremen Reihe A 2,3:141-148

Kodric-Brown A (1977) Reproductive success and the evolution of breeding territories in pupfish (*Cyprinodon*). Evolution 31:750-766

Kortmulder K (1972) A comparative study in colour patterns and behaviour in seven Asiatic *Barbus* species (Cyprinidae, Ostariophysi, Osteichthyes). A progress report. Behaviour Suppl 19:1-331

Kortmulder K, Feldbrugge EJ, De Silva SS (1978) A combined field study of *Barbus* (= *Puntius*) *nigrofasciatus* Günther (Pisces, Cyprinidae) and water chemistry of its habitat in Sri Lanka. Neth J Zool 28:111-131

Kramer DL (1973) Parental behaviour in the blue gourami *Trichogaster trichopterus* (Pisces, Belontiidae) and its induction during exposure to varying numbers of conspecific eggs. Behaviour 47:14-32

Krekorian, C O'Neil (1976) Field observations in Guyana on the reproductive biology of the spraying characid, *Copeina arnoldi* Regan. Am Midl Nat 96:88-97

Krekorian, C O'Neil, Dunham DW (1972a) Preliminary observations on the reproductive and parental behavior of the spraying characid *Copeina arnoldi* Regan. Z Tierpsychol 31:419-437

Krekorian, C O'Neil, Dunham DW (1972b) Parental egg care in the spraying characid (*Copeina arnoldi* Regan): role of the spawning surface. Anim Behav 20:356-360

Krischik VA, Weber PG (1975) Induced parental care in male convict cichlid fish. Dev Psychobiol 8:1-11

Kruuk H (1963) Diurnal periodicity in the activity of the common sole, *Solea vulgaris* Quensel. Neth J Sea Res 2:1-28

Kuenzer P (1966) Wie "erkennen" junge Buntbarsche ihre Eltern? Umschau Wiss Tech 24:795-800

Kuenzer P (1968) Die Auslösung der Nachfolgereaktion bei erfahrungslosen Jungfischen von *Nannacara anomala* (Cichlidae). Z Tierpsychol 25:257-314

Kuenzer P (1975) Analyse der auslösenden Reizsituationen für die Anschwimm-, Eindring- und Fluchtreaktion junger *Hemihaplochromis multicolor* (Cichlidae). Z Tierpsychol 38:505-545

Kuenzer E, Kuenzer P (1962) Untersuchungen zur Brutpflege der Zwergcichliden *Apistogramma reitzigi* and *A borellii*. Z Tierpsychol 19:56-83

Kuenzer P, Peters L (1974) Versuche zur Auslösung und Umstellung der Brutpflegephasen bei ♀♀ von *Nannacara anomala* (Cichlidae). Z Tierpsychol 35:425-436

Kühme W (1962) Das Schwarmverhalten elterngeführter Jungcichliden (Pisces). Z Tierpsychol 19:513-518

Kühme W (1963) Chemisch ausgelöste Brutpflege- und Schwarmreaktionen bei *Hemichromis bimaculatus* (Pisces). Z Tierpsychol 20:688-704

Kummer H (1968) Social organization of Hamadryas baboons. Univ Chicago Press, Chicago, p 189

Lagler KF, Bardach JE, Miller RR (1962) Ichthyology. Wiley, New York, p 545

Langescheid C (1968) Vergleichende Untersuchungen über die angeborene Größenunterscheidung bei *Tilapia nilotica* und *Hemihaplochromis multicolor* (Pisces: Cichlidae). Experientia 24:963-964

Larsen GL (1976) Social behavior and feeding ability of two phenotypes of *Gasterosteus aculeatus* in relation to their spatial and trophic segregation in a temperate lake. Can J Zool 54:107-121

Leathers AL (1911) A biological survey of the sand dune region of the south shore, Saginaw Bay, Michigan. Mich Geo Bio Surv Pub 4, Biol Serv 2:243-255

Lee G (1937) Oral gestation in the marine catfish, *Galeichthys felis*. Copeia 1937:49-56

Leim AH, Scott WB (1966) Fishes of the Atlantic coast of Canada. Bull Fish Res Board Can 155:1-485

Lennon RE (1954) Feeding mechanism of the sea lamprey and its effect on host fishes. US Fish Wildl Serv Fish Bull 56:246-293

Leong RJH, O'Connell CP (1969) A laboratory study of particulate and filter feeding of the northern anchovy (*Engraulis mordax*). J Fish Res Board Can 26:557-582

Liem KF, Osse JWM (1975) Biological versatility, evolution, and food resource exploitation in African cichlid fishes. Am Zool 15:427-454

Liem KF, Stewart DJ (1976) Evolution of the scale-eating cichlid fishes of Lake Tanganyika: a generic revision with a description of a new species. Bull Mus Comp Zool Harv Univ 147:319-350

Liley NR (1966) Ethological isolating mechanisms in four sympatric species of poeciliid fishes. Behaviour Suppl 13:1-197

Limbaugh C (1964) Notes on the life history of two California pomacentrids: garibaldis, *Hypsypops rubicunda* (Girard), and blacksmiths, *Chromis punctipinnis* (Cooper). Pac Sci 18:41-50

Lindsey CC (1978) Form, function, and locomotory habits in fish. In: Hoar WS, Randall DJ (eds) Fish physiology, vol 7. Academic, New York, pp 1-100

Lineaweaver TH, Backus RH (1970) The natural history of sharks. Lippincott, Philadelphia, p 256

Loftus KH (1958) Studies on river-spawning populations of lake trout in eastern Lake Superior. Trans Am Fish Soc 87:259-277

Lohnisky K (1966) The spawning behaviour of the brook lamprey, *Lampetra planeri* (Bloch, 1784). Vestn Cesk Spol Zool 30:289-307

Losey GS (1972) Predation protection in the poison-fang blenny, *Meiacanthus atrodorsalis*, and its mimics, *Ecsenius bicolor* and *Runula laudandus* (Blenniidae). Pac Sci 26:129-139

Low RM (1971) Interspecific territoriality in a pomacentrid reef fish, *Pomacentrus flavicauda* Whitley. Ecology 52:648-654

Lowe-McConnell RH (1956) The breeding behaviour of *Tilapia* species (Pisces: Cichlidae) in natural waters: observations on *T karomo* Poll and *T variabilis* Boulenger. Behaviour 9:140-163

Lowe-McConnell RH (1959) Breeding behaviour patterns and ecological differences between *Tilapia* species and their significance for evolution within the genus *Tilapia* (Pisces: Cichlidae). Proc Zool Soc Lond 132:1-30

Lowe-McConnell RH (1964) The fishes of the Rupununi savanna district of British Guiana, South America, Part I. Ecological groupings of fish species and effects of the seasonal cycle on the fish. J Linn Soc Lond Zool 45:103-144

Lowe-McConnell RH (1969) Speciation in tropical freshwater fishes. Biol J Linn Soc 1:51-75

Lüling KH (1963) The archer fish. Sci Am 209(1):100-108

Lüling KH (1969) Das Beutespucken von Schützenfisch *Toxotes jaculatrix* und Zwergfadenfisch *Colisa lalia*. Bonn Zool Beitr 20:416-422

Machemer L (1970) Qualitative und quantitative Verhaltensbeobachtungen an Paradiesfisch-♂♂, *Macropodus opercularis* L (Anabantidae, Teleostei). Z Tierpsychol 27:563-590

MacNae W (1968) A general account of the fauna and flora of mangrove swamps and forests in the Indo-West-Pacific region. Adv Mar Biol 6:73-270

Magnus DBE (1967a) Zur Ökologie sedimentbewohnender *Alpheus*-Garnelen (Decapoda, Natantia) des Roten Meeres. Helgol Wiss Meeresunters 15:506-522

Magnus DBE (1967b) Zur Deutung der Igelstellung beim Jungfischschwarm des Korallenwelses, *Plotosus anguillaris* (Bloch) (Pisces, Nematognathi, Plotosidae) im Biotop. Verh Dtsch Zool Ges 4:402-409

Major PF (1973) Scale feeding behavior of the leatherjacket, *Scomberoides lysan* and two species of the genus *Oligoplites* (Pisces: Carangidae). Copeia 1973:151-154

Malinin LK (1969) Uchastki obitaniya i instinkt vozvrashcheniya ryb (Home range and homing instinct of fish). Zool Zh 48:381-391, Fish Res Board Can, Transl Ser 2050

Mansueti R (1963) Symbiotic behavior between small fishes and jellyfishes, with new data on that between the stromateid, *Peprilus alepidotus*, and the scyphomedusa *Chrysaora quinquecirrha*. Copeia 1963:40-80

Mapstone GM, Wood EM (1975) The ethology of *Abudefduf luridus* and *Chromis chromis* (Pisces: Pomacentridae) from the Azores. J Zool Lond 175:179-199

Mariscal R (1972) Behavior of symbiotic fishes and sea anemones. In: Winn HE, Olla BL (eds) Behavior of marine animals. Plenum, London, pp 327-360

Markl H (1968) Das Schutzverhalten eines Welses (*Hassar orestis* Steindachner) gegen Angriffe von Piranhas (*Serrasalmus nattereri* Kner). Z Tierpsychol 26:385-389

Marshall NB (1965) The life of fishes. Weidenfeld and Nicolson, London, p 402

Marshall TC (1964) Fishes of the Great Barrier Reef. Angus and Robertson, Sydney, p 566

Martin NV (1957) Reproduction of lake trout in Algonquin Park, Ontario. Trans Am Fish Soc 86:231-244

Mason JC, Chapman DW (1965) Significance of early emergence, environmental rearing capacity, and behavioral ecology of juvenile coho salmon in stream channels. J Fish Res Board Can 22:173-190

Mathisen OA (1962) The effect of altered sex ratios on the spawning of red salmon. Univ Wash Publ Fish New Ser 1:139-245

Matthes H (1961) Feeding habit of some central African freshwater fishes. Nature (London) 192:78-80

Mayr E (1963) Animal species and evolution. Belknap Harvard, Cambridge, p 797

McAllister DE (1967) The significance of ventral bioluminescence in fishes. J Fish Res Board Can 24:537-554

McCart PJ (1969) Digging behaviour of *Oncorhynchus nerka* spawning in streams at Babine Lake, British Columbia. In: Northcote TG (ed) Symposium on salmon and trout in streams. Univ B C Vancouver, pp 39-51

McCart PJ (1970) A polymorphic population of *Oncorhynchus nerka* at Babine Lake, B C, involving anadromous (sockeye) and non-anadromous (kokanee) forms. Ph D Thesis, Univ B C, Vancouver, p 135

McCrimmon HR (1968) Carp in Canada. Bull Fish Res Board Can 165:1-93

McInerney JE (1969) Reproductive behaviour of the blackspotted stickleback, *Gasterosteus wheatlandi*. J Fish Res Board Can 26:2061-2075

McIntyre JD (1969) Spawning behavior of the brook lamprey, *Lampetra planeri*. J Fish Res Board Can 26:3252-3254

McKaye KR (1977) Competition for breeding sites between the cichlid fishes of Lake Jiloa, Nicaragua. Ecology 58:291-302

McKaye KR, Barlow GW (1976a) Competition between color morphs of the Midas cichlid, *Cichlasoma citrinellum*, in Lake Jiloa, Nicaragua. In: Thorson TB (ed) Investigations of the ichthyofauna of Nicaraguan lakes. Univ Nebraska, Lincoln, pp 465-475

McKaye KR, Barlow GW (1976b) Chemical recognition of young by the Midas cichlid, *Cichlasoma citrinellum*. Copeia 1976:276-282

McKaye KR, McKaye NM (1977) Communal care and kidnapping of young by parental cichlids. Evolution 31:674-681

McKenzie JA (1964) The reproductive behaviour of the brook stickleback, *Eucalia inconstans* (Kirtland). Ph D Thesis, Univ Western Ontario, London p 135

McKenzie JA (1969) The courtship behavior of the male brook stickleback, *Culaea inconstans* (Kirtland). Can J Zool 47:1281-1286

McKenzie JA (1974) The parental behavior of the male brook stickleback, *Culaea inconstans* (Kirtland). Can J Zool 52:649-652

McKenzie JA, Keenleyside MHA (1970) Reproductive behavior of ninespine sticklebacks (*Pungitius pungitius* (L)) in South Bay, Manitoulin Island, Ontario. Can J Zool 48:55-61

McPhail JD (1963) Geographic variation in North American ninespine sticklebacks, *Pungitius pungitius*. J Fish Res Board Can 20:27-44

McPhail JD, Lindsey CC (1970) Freshwater fishes of northwestern Canada and Alaska. Bull Fish Res Board Can 173:1-381

Mertz JC, Barlow GW (1966) On the reproductive behavior of *Jordanella floridae* (Pisces: Cyprinodontidae) with special reference to a quantitative analysis of parental fanning. Z Tierpsychol 23:537-554

Miller HC (1963) The behavior of the pumpkinseed sunfish, *Lepomis gibbosus* (Linnaeus), with notes on the behavior of other species of *Lepomis* and the pigmy sunfish, *Elassoma evergladei*. Behaviour 22:88-151

Miller RJ (1964) Studies on the social behavior of the blue gourami, *Trichogaster trichopterus* (Pisces: Belontiidae). Copeia 1964:469-496

Miller RJ, Robison HW (1974) Reproductive behavior and phylogeny in the genus *Trichogaster* (Pisces: Anabantoidei). Z Tierpsychol 34:484-499

Miller RR (1966) Geographical distribution of Central American freshwater fishes. Copeia 1966:773-802

Minckley CO, Klaassen HE (1969) Burying behavior of the plains killifish, *Fundulus kansae*. Copeia 1969:200-201

Mok H-K (1978) Scale-feeding in *Tydemania navigatoris* (Pisces: Triacanthodidae). Copeia 1978:338-340

Morris D (1954) The reproductive behaviour of the river bullhead (*Cottus gobio* L), with special reference to the fanning activity. Behaviour 7:1-32

Morris D (1956) The function and causation of courtship ceremonies. In: Grassé PP (ed) L'instinct dans le comportement des animaux et de l'homme. Masson, Paris, pp 261-286

Morris D (1958) The reproductive behaviour of the ten-spined stickleback (*Pygosteus pungitius* L). Behaviour Suppl 6:1-154

Mortimer MAE (1960) Observations on the biology of *Tilapia andersonii* (Castelnau) (Pisces: Cichlidae), in northern Rhodesia. Rep J Fish Res Org North Rhod 9:42-67

Mount DI (1959) Spawning behavior of the bluebreast darter, *Etheostoma camurum* (Cope). Copeia 1959:240-243

Moyer JT (1975) Reproductive behavior of the damselfish *Pomacentrus nagasakiensis* at Miyake-jima, Japan. Jpn J Ichthyol 22:151-163

Moyer JT (1976) Geographical variation and social dominance in Japanese populations of the anemonefish *Amphiprion clarkii*. Jpn J Ichthyol 23:12-22

Moyer JT, Bell LJ (1976) Reproductive behavior of the anemonefish *Amphiprion clarkii* at Miyake-jima, Japan. Jpn J Ichthyol 23:23-32

Moyer JT, Nakazono A (1978) Protandrous hermaphroditism in six species of the anemonefish genus *Amphiprion* in Japan. Jpn J Ichthyol 25:101-106

Moyer JT, Sawyers CE (1973) Territorial behavior of the anemonefish *Amphiprion xanthurus* with notes on the life history. Jpn J Ichthyol 20:85-93

Munro ISR (1967) The fishes of New Guinea. Dep Agric Stock Fish. Port Moresby, New Guinea, p 651

Myers GS (1939) A possible method of evolution of oral brooding habits in cichlid fishes. Stanford Ichthyol Bull 1:85-87

Myers GS (1950) Supplementary notes on the flying characid fishes, especially *Carnegiella*. Stanford Ichthyol Bull 3:182-183

Myers GS (1966) Derivation of the freshwater fish fauna of Central America. Copeia 1966:766-773

Myrberg AA (1964) An analysis of the preferential care of eggs and young by adult cichlid fishes. Z Tierpsychol 21:53-98

Myrberg AA (1965) A descriptive analysis of the behaviour of the African cichlid fish, *Pelmatochromis guentheri* (Sauvage). Anim Behav 13:312-329

Myrberg AA (1966) Parental recognition of young in cichlid fishes. Anim Behav 14:565-571

Myrberg AA (1972) Ethology of the bicolor damselfish, *Eupomacentrus partitus* (Pisces: Pomacentridae): A comparative analysis of laboratory and field behaviour. Anim Behav Monogr 5:197-283

Myrberg AA (1973) Underwater television — a tool for the marine biologist. Bull Mar Sci 23:824-836

Myrberg AA (1975) The role of chemical and visual stimuli in the preferential discrimination of young by the cichlid fish *Cichlasoma nigrofasciatum* (Günther). Z Tierpsychol 37:274-297

Myrberg AA, Brahy BD, Emery AR (1967) Field observations on reproduction of the damselfish, *Chromis multilineata* (Pomacentridae) with additional notes on general behavior. Copeia 1967:819-827

190

Myrberg AA, Kramer E, Heinecke P (1965) Sound production by cichlid fishes. Science 149:555-558

Myrberg AA, Spires JY (1972) Sound discrimination by the bicolor damselfish, *Eupomacentrus partitus*. J Exp Biol 57:727-735

Myrberg AA, Ha SJ, Walewski S, Banbury JC (1972) Effectiveness of acoustic signals in attracting epipelagic sharks to an underwater sound source. Bull Mar Sci 22:926-949

Needham PR (1961) Observations on the natural spawning of eastern brook trout. Calif Fish Game 47:27-40

Needham PR, Taft AC (1934) Observations on the spawning of steelhead trout. Trans Am Fish Soc 64:332-338

Needham PR, Vaughan TM (1952) Spawning of the Dolly Varden, *Salvelinus malma*, in Twin Creek, Idaho. Copeia 1952:197-199

Neill SRStJ, Cullen JM (1974) Experiments on whether schooling by their prey affects the hunting behaviour of cephalopods and fish predators. J Zool Lond 172:549-569

Nelson JS (1976) Fishes of the world. Wiley, New York, p 416

Nelson K (1965) After-effects of courtship in the male three-spined stickleback. Z Vergl Physiol 50:569-597

New JG (1966) Reproductive behavior of the shield darter, *Percina peltata peltata*, in New York. Copeia 1966:20-28

Newman HH (1907) Spawning behavior and sexual dimorphism in *Fundulus heteroclitus* and allied fish. Biol Bull 12:314-348

Newman MA (1956) Social behavior and interspecific competition in two trout species. Physiol Zool 29:64-81

Noakes DLG (1973) Parental behavior and some histological features of scales in *Cichlasoma citrinellum* (Pisces: Cichlidae). Can J Zool 51:619-622

Noakes DLG, Barlow GW (1973a) Ontogeny of parent-contacting in young *Cichlasoma citrinellum* (Pisces: Cichlidae). Behaviour 46:221-255

Noakes DLG, Barlow GW (1973b) Cross-fostering and parent–offspring responses in *Cichlasoma citrinellum* (Pisces: Cichlidae). Z Tierpsychol 33:147-152

Noble GK (1939) The role of dominance in the social life of birds. Auk 56:263-273

Noble GK, Curtis B (1939) The social behavior of the jewel fish, *Hemichromis bimaculatus* Gill. Bull Am Mus Nat Hist 76:1-46

Norman JR, Greenwood PH (1975) A history of fishes, 3rd edn. Benn, London, p 467

Nursall JR (1973) Some behavioral interactions of spottail shiners (*Notropis hudsonius*), yellow perch (*Perca flavescens*), and northern pike (*Esox lucius*). J Fish Res Board Can 30:1161-1178

Nursall JR (1977) Territoriality in redlip blennies (*Ophioblennius atlanticus* — Pisces: Blenniidae). J Zool Lond 182:205-223

Nyman K-J (1953) Observations on the behaviour of *Gobius microps*. Acta Soc Fauna Flora Fenn 69(5):1-11

O'Connell CP (1972) The interrelation of biting and filtering in the feeding activity of the northern anchovy (*Engraulis mordax*). J Fish Res Board Can 29:285-293

Odum EP, de la Cruz AA (1963) Detritus as a major component of ecosystems. AIBS Bull 13:39-40

Odum WE (1970) Utilization of the direct grazing and plant detritus food chains by the striped mullet *Mugil cephalus*. In: Steele JH (ed) Marine food chains. Oliver and Boyd, Edinburgh, pp 222-240

Ogden JC, Buckman NS (1973) Movements, foraging groups, and diurnal migrations of the striped parrotfish *Scarus croicensis* Bloch (Scaridae). Ecology 54:589-596

Olla BL, Bejda AJ, Martin AD (1974) Daily activity, movements, feeding, and seasonal occurrence in the tautog, *Tautoga onitis*. Fish Bull 72:27-35

Olla BL, Katz HM, Studholme AL (1970) Prey capture and feeding motivation in the bluefish, *Pomatomus saltatrix*. Copeia 1970:360-362

Olla BL, Samet CE, Studholme AL (1972) Activity and feeding behavior of the summer flounder (*Paralichthys dentatus*) under controlled laboratory conditions. Fish Bull 70:1127-1136

Ono Y, Uematsu T (1957) Mating ethogram in *Oryzias latipes*. J Fac Sci Hokkaido Univ Ser 6 13:197-202

Oppenheimer JR (1970) Mouthbreeding in fishes. Anim Behav 18:493-503

Oppenheimer JR, Barlow GW (1968) Dynamics of parental behavior in the black-chinned mouthbreeder, *Tilapia melanotheron* (Pisces: Cichlidae). Z Tierpsychol 25:889-914

191

Orton JH, Jones JW, King GM (1938) The male sexual stage in salmon parr (*Salmo salar* L juv). Proc R Soc Lond B 125:103-114

Outram DN, Humphreys RD (1974) The Pacific herring in British Columbia waters. Fish Mar Serv Nanaimo, B C Circ 100:1-26

Parr AE (1927) A contribution to the theoretical analysis of the schooling behavior of fishes. Occas Pap Bingham Oceanogr Coll 1:1-32

Parrish RH (1972) Symbiosis in the blacktail snailfish, *Careproctus melanurus*, and the box crab, *Lopholithodes foraminatus*. Calif Fish Game 58:239

Pearson NE (1937) The fishes of the Beni-Mamoré and Paraguay basins, and a discussion of the origin of the Paraguayan fauna. Proc Calif Acad Sci 23:99-114

Peden AE (1973) Variation in anal spot expression of gambusiin females and its effect on male courtship. Copeia 1973:250-263

Peden AE, Corbett CA (1973) Commensalism between a liparid fish, *Careproctus* sp, and the lithodid box crab, *Lopholithodes foraminatus*. Can J Zool 51:555-556

Pelkwijk JJ ter, Tinbergen N (1937) Eine reizbiologische Analyse einiger Verhaltensweisen von *Gasterosteus aculeatus* L. Z Tierpsychol 1:193-200

Peters HM (1965) Über larvale Haftorgane bei *Tilapia* (Cichlidae, Teleostei) und ihre Rückbildung in der Evolution. Zool Jahrb Physiol 71:287-300

Pfeiffer W (1962) The fright reaction of fish. Biol Rev 37:495-511

Pfeiffer W (1963) The fright reaction of North American fish. Can J Zool 41:69-77

Pfeiffer W (1967) Schreckreaktion und Schreckstoffzellen bei Ostariophysi und Gonorhynchiformes. Z Vergl Physiol 56:380-396

Pfeiffer W (1977) The distribution of fright reaction and alarm substance cells in fishes. Copeia 1977:653-665

Pflieger WL (1965) Reproductive behavior of the minnows, *Notropis spilopterus* and *Notropis whipplii*. Copeia 1965:1-8

Phillips RR (1971) The relationship between social behavior and the use of space in the benthic fish *Chasmodes bosquianus* Lacépède (Teleostei, Blenniidae). II. The effect of prior residency on social and enclosure behavior. Z Tierpsychol 29:398-408

Phillips RR (1977) Behavioral field study of the Hawaiian rockskipper, *Istiblennius zebra* (Teleostei, Blenniidae). I. Ethogram. Z Tierpsychol 43:1-22

Pike GC (1951) Lamprey marks on whales. J Fish Res Board Can 8:275-280

Polder JJW (1971) On gonads and reproductive behaviour in the cichlid fish *Aequidens portalegrensis* (Hensel). Neth J Zool 21:265-365

Pollard RA (1955) Measuring seepage through salmon spawning gravel. J Fish Res Board Can 12:706-741

Popper D, Fishelson L (1973) Ecology and behavior of *Anthias squamipinnis* (Peters, 1855) (Anthiidae, Teleostei) in the coral habitat of Eilat (Red Sea). J Exp Zool 184:409-423

Potts GW (1970) The schooling ethology of *Lutianus monostigma* (Pisces) in the shallow reef environment of Aldabra. J Zool Lond 161:223-235

Potts GW (1974) The colouration and its behavioural significance in the corkwing wrasse, *Crenilabrus melops*. J Mar Biol Assoc U K 54:925-938

Powles PM (1958) Studies of reproduction and feeding of Atlantic cod (*Gadus callarias* L) in the southwestern Gulf of St Lawrence. J Fish Res Board Can 15:1383-1402

Pyke GH, Pulliam HR, Charnov EL (1977) Optimal foraging: a selective review of theory and tests. Q Rev Biol 52:137-154

Qasim SZ (1956) The spawning habits and embryonic development of the shanny (*Blennius pholis* L). Proc Zool Soc Lond 127:79-93

Quertermus CJ, Ward JA (1969) Development and significance of two motor patterns used in contacting parents by young orange chromides (*Etroplus maculatus*). Anim Behav 17:624-635

Radakov DV (1973) Schooling in the ecology of fish. Wiley, New York, p 173

Randall JE (1956) A revision of the surgeon fish genus *Acanthurus*. Pac Sci 10:159-235

Randall JE (1961a) A contribution to the biology of the convict surgeonfish of the Hawaiian Islands, *Acanthurus triostegus sandvicensis*. Pac Sci 15:215-272

Randall JE (1961b) Observations on the spawning of surgeonfishes (Acanthuridae) in the Society Islands. Copeia 1961:237-238

Randall JE (1967) Food habits of reef fishes of the West Indies. Stud Trop Oceanogr 5:665-847

Randall JE (1968) Caribbean reef fishes. T F H, Jersey City, p 318

Randall JE, Randall HA (1960) Examples of mimicry and protective resemblance in tropical marine fishes. Bull Mar Sci Gulf Caribb 10:444-480

Randall JE, Randall HA (1963) The spawning and early development of the Atlantic parrotfish, *Sparisoma rubripinne*, with notes on other scarid and labrid fishes. Zoologica 48:49-60

Raney EC (1939) The breeding habits of the silvery minnow, *Hybognathus regius* Girard. Am Midl Nat 21:674-680

Raney EC (1940a) Nests under the water. Bull NY Zool Soc 43:127-135

Raney EC (1940b) The breeding behavior of the common shiner, *Notropis cornutus* (Mitchill). Zoologica 25:1-14

Raney EC, Backus RH, Crawford RW, Robins CR (1953) Reproductive behavior in *Cyprinidon variegatus* Lacépède, in Florida. Zoologica 38:97-104

Rasa OAE (1969) Territoriality and the establishment of dominance by means of visual cues in *Pomacentrus jenkinsi* (Pisces: Pomacentridae). Z Tierpsychol 26:825-845

Reese ES (1964) Ethology and marine zoology. Oceanogr Mar Biol 2:455-488

Reese ES (1973) Duration of residence by coral reef fishes on "home" reefs. Copeia 1973:145-149

Reese ES (1975) A comparative field study of the social behavior and related ecology of reef fishes of the family Chaetodontidae. Z Tierpsychol 37:37-61

Reese ES (1977) Coevolution of corals and coral feeding fishes of the family Chaetodontidae. Proc 3rd Int Coral Reef Symp Univ Miami, Florida, pp 267-274

Reeves CD (1907) The breeding habits of the rainbow darter (*Etheostoma coeruleum* Storer), a study in sexual selection. Biol Bull 14:35-59

Reid MJ, Atz JW (1958) Oral incubation in the cichlid fish *Geophagus jurupari* Heckel. Zoologica 43:77-88

Reighard JE (1910) Methods of studying the habits of fishes, with an account of the breeding habits of the horned dace. Bull US Bur Fish XVII(2):1111-1136

Reighard JE (1943) The breeding habits of the river chub *Nocomis micropogon* (Cope). Pap Mich Acad Sci Arts Lett 28:397-423

Reimers PE (1968) Social behavior among juvenile fall chinook salmon. J Fish Res Board Can 25:2005-2008

Reinboth R (1970) Intersexuality in fishes. Mem Soc Endocrinol 18:515-543

Reinboth R (1973) Dualistic reproductive behavior in the protogynous wrasse *Thalassoma bifasciatum* and some observations on its day-night changeover. Helgol Wiss Meeresunters 24:174-191

Reisman HM, Cade TJ (1967) Physiological and behavioral aspects of reproduction in the brook stickleback, *Culaea inconstans*. Am Midl Nat 77:257-295

Ribbink AJ (1971) The behaviour of *Hemihaplochromis philander*, a South African cichlid fish. Zool Afr 6:263-288

Ribbink AJ (1977) Cuckoo among Lake Malawi cichlid fish. Nature (London) 267:243-244

Richardson LR (1939) The spawning behavior of *Fundulus diaphanus* (Le Sueur). Copeia 1939:165-167

Roberts NJ, Winn HE (1962) Utilization of the senses in feeding behavior of the Johnny darter, *Etheostoma nigrum*. Copeia 1962:567-570

Roberts TR (1970) Scale-eating American characoid fishes, with special reference to *Probolodus heterostomus*. Proc Calif Acad Sci 38:383-390

Roberts TR (1972) Ecology of fishes in the Amazon and Congo basins. Bull Mus Comp Zool Harv Univ 143:117-147

Robertson DR (1972) Social control of sex reversal in a coral-reef fish. Science 177:1007-1009

Robertson DR (1973) Field observations on the reproductive behaviour of a pomacentrid fish, *Acanthochromis polyacanthus*. Z Tierpsychol 32:319-324

Robertson DR, Choat JH (1974) Protogynous hermaphroditism and social systems in labrid fish. Proc 2nd Int Coral Reef Symp Brisbane 1:217-225

Robertson DR, Hoffman SG (1977) The roles of female mate choice and predation in the mating systems of some tropical labroid fishes. Z Tierpsychol 45:298-320

Robertson DR, Sweatman HPA, Fletcher EA, Cleland MG (1976) Schooling as a mechanism for circumventing the territoriality of competitors. Ecology 57:1208-1220

Robison HW (1975) A qualitative analysis of courtship and reproductive behavior in the anabantoid fish *Trichogaster pectoralis* (Regan) (Pisces: Anabantoidei). Proc Okla Acad Sci 55:65-71

Rosen DE, Gordon M (1953) Functional anatomy and evolution of male genitalia in poeciliid fishes. Zoologica 38:1-48

Rosen DE, Tucker A (1961) Evolution of secondary sexual characters and sexual behavior patterns in a family of viviparous fishes (Cyprinodontiformes: Poeciliidae). Copeia 1961:201-212

Rosen MW, Cornford NE (1971) Fluid friction of fish slimes. Nature (London) 234:49-51

Ross RM (1978) Territorial behavior and ecology of the anemonefish *Amphiprion melanopus* on Guam. Z Tierpsychol 46:71-83

Rounsefell GA (1930) Contribution to the biology of the Pacific herring, *Clupea pallasii*, and the condition of the fishery in Alaska. Bull US Bur Fish 45:227-320

Rowland WJ (1974a) Reproductive behavior of the fourspine stickleback, *Apeltes quadracus*. Copeia 1974:183-194

Rowland WJ (1974b) Ground nest construction in the fourspine stickleback, *Apeltes quadracus*. Copeia 1974:788-789

Royce WF (1951) Breeding habits of lake trout in New York. US Fish Wildl Serv Fish Bull 52:59-76

Russell BC (1971) Underwater observations on the reproductive activity of the demoiselle *Chromis dispilus* (Pisces: Pomacentridae). Mar Biol 10:22-29

Ruwet JC (1962) La reproduction des *Tilapia macrochir* (Blgr) et *melanopleura* (Dum) (Pisces: Cichlidae) au lac barrage de la Lufira (Haut-Katanga). Rev Zool Bot Afr 66:243-271

Ruwet JC (1963) Observations sur le comportement sexuel de *Tilapia macrochir* Blgr (Pisces: Cichlidae) au lac de retenue de la Lufira (Katanga). Behaviour 20:242-250

Sale PF (1970) Behaviour of the Humbug fish. Aust Nat Hist 16:362-366

Sale PF (1971a) The reproductive behaviour of the pomacentrid fish, *Chromis caeruleus*. Z Tierpsychol 29:156-164

Sale PF (1971b) Extremely limited home range in a coral reef fish, *Dascyllus aruanus* (Pisces: Pomacentridae). Copeia 1971:324-327

Sale PF (1972) Influence of corals in the dispersion of the pomacentrid fish, *Dascyllus aruanus*. Ecology 53:741-744

Sale PF (1974) Mechanisms of co-existence in a guild of territorial fishes at Heron Island. Proc 2nd Int Coral Reef Symp 1:193-206

Sale PF (1975) Patterns of use of space in a guild of territorial reef fishes. Mar Biol 29:89-97

Sale PF (1977) Maintenance of high diversity in coral reef fish communities. Am Nat 111:337-359

Sartori JD, Bright TJ (1973) Hydrophonic study of the feeding activities of certain Bahamian parrotfishes, family Scaridae. Hydrolab J 2:25-56

Savage T (1963) Reproductive behavior of the mottled sculpin, *Cottus bairdi* Girard. Copeia 1963:317-325

Scalet CG (1973) Reproduction of the orangebelly darter, *Etheostoma radiosum cyanorum* (Osteichthyes: Percidae). Am Midl Nat 89:156-165

Schaefer MB (1937) Notes on the spawning of the Pacific herring, *Clupea pallasii*. Copeia 1937:57

Schoener TW (1971) Theory of feeding strategies. Ann Rev Ecol Syst 2:369-404

Schultz LP (1935) The spawning habits of the chub, *Mylocheilus caurinus* — a forage fish of some value. Trans Am Fish Soc 65:143-147

Schutz DC, Northcote TG (1972) An experimental study of feeding behavior and interaction of coastal cutthroat trout (*Salmo clarki clarki*) and Dolly Varden (*Salvelinus malma*). J Fish Res Board Can 29:555-565

Schutz F (1956) Vergleichende Untersuchungen über die Schreckreaktion bei Fischen und deren Verbreitung. Z Vergl Physiol 38:84-135

Schwarz A (1974) The inhibition of aggressive behavior by sound in the cichlid fish, *Cichlasoma centrarchus*. Z Tierpsychol 35:508-517

Scott WB, Crossman EJ (1973) Freshwater fishes of Canada. Bull Fish Res Board Can 184:1-966

Seghers BH (1974a) Schooling behavior in the guppy (*Poecilia reticulata*): an evolutionary response to predation. Evolution 28:486-489

Seghers BH (1974b) Geographic variation in the responses of guppies (*Poecilia reticulata*) to aerial predators. Oecologia 14:93-98

Sevenster P (1951) The mating of the sea stickleback. Discovery 12:52-56

Sevenster P (1961) A causal analysis of a displacement activity (fanning in *Gasterosteus aculeatus* L). Behaviour Suppl 9:1-170

Shallenberger RJ, Madden WD (1973) Luring behavior in the scorpionfish, *Iracundus signifer*. Behaviour 47:33-47

194

Shaw E (1962) The schooling of fishes. Sci Am 206(6):128-138

Shaw E (1970) Schooling in fishes: critique and review. In: Aronson LR, Tobach E, Lehrman DS, Rosenblatt JS (eds) Development and evolution of behavior. Freeman, San Francisco, pp 452-480

Shaw E (1978) Schooling fishes. Am Sci 66:166-175

Shaw ES, Aronson LR (1954) Oral incubation in *Tilapia macrocephala*. Bull Am Mus Nat Hist 103:375-416

Shuleikin VV (1958) How the pilot fish moves with the speed of the shark. Dokl Akad Nauk SSSR 119:140-143

Sjolander S (1972) Feldbeobachtungen an einigen westafrikanischen Cichliden. Aquar Terr 19:42-45

Slaney PA, Northcote TG (1974) Effects of prey abundance on density and territorial behavior of young rainbow trout (*Salmo gairdneri*) in laboratory stream channels. J Fish Res Board Can 31:1201-1209

Smith BR (1971) Sea lampreys in the Great Lakes of North America. In: Hardisty MW, Potter IC (eds) The biology of lampreys, vol 1. Academic, New York, pp 207-247

Smith OR (1941) The spawning habits of cutthroat and eastern brook trouts. J Wildl Manage 5:461-471

Smith RJF (1970) Control of prespawning behavior of sunfish (*Lepomis gibbosus* and *Lepomis megalotis*). II. Environmental factors. Anim Behav 18:575-587

Smith-Grayton PK, Keenleyside MHA (1978) Male–female parental roles in *Herotilapia multispinosa* (Pisces: Cichlidae). Anim Behav 26:520-526

Sprenger K (1971) The red hump *Geophagus*. Buntbarsche Bull 26:24-25

Springer S (1957) Some observations on the behavior of schools of fishes in the Gulf of Mexico and adjacent waters. Ecology 38:166-171

Springer S (1967) Social organization of shark populations. In: Gilbert PW, Mathewson RF, Rall DP (eds) Sharks, skates, and rays. Johns Hopkins Press, Baltimore, pp 149-174

Springer VG, Smith-Vaniz WF (1972) Mimetic relationships involving fishes of the family Blenniidae. Smithson Contrib Zool 112:1-36

Starck WA, Davis WP (1966) Night habits of fishes of Alligator Reef, Florida. Ichthyologica 38:313-356

Stasko AB (1975) Underwater biotelemetry, an annotated bibliography. Can Fish Mar Serv Tech Rep 534:1-31

Stebbins RC, Kalk M (1961) Observations on the natural history of the mud-skipper, *Periophthalmus sobrinus*. Copeia 1961:18-27

Steinitz H (1959) Observations on *Pterois volitans* (L) and its venom. Copeia 1959:158-160

Stephens JS, Johnson RK, Key GS, McCosker JE (1970) The comparative ecology of three sympatric species of California blennies of the genus *Hypsoblennius* Gill (Teleostomi, Blenniidae). Ecol Monogr 40:213-233

Stephenson W, Searles RB (1960) Experimental studies on the ecology of intertidal environments at Heron Island. I. Exclusion of fish from beach rock. Aust J Mar Freshwater Res 11:241-267

Sterba G (1962) Die Neunaugen (Petromyzonidae). Handb Binnenfisch Mitteleur 36:263-352

Sterba G (1966) Freshwater fishes of the World. Revised English ed. Studio Vista, London, p 877

Strahan R (1963a) The behaviour of myxinoids. Acta Zool 44:73-102

Strahan R (1963b) The behaviour of Myxine and other myxinoids. In: Brodal A, Fänge R (eds) The biology of Myxine. Scand Univ Books, Oslo, pp 22-32

Strasburg DW, Jones EC, Iverson RTB (1968) Use of a small submarine for biological and oceanographic research. J Conseil 31:410-426

Struhsaker TT (1969) Correlates of ecology and social organization among African cercopithecines. Folia Primatol 11:80-118

Sumner FB (1911) The adjustment of flatfishes to various backgrounds: a study of adaptive color change. J Exp Zool 10:409-505

Sumner FB (1934) Does "protective coloration" protect? — Results of some experiments with fishes and birds. Proc Natl Acad Sci USA 20:559-564

Sumner FB (1935) Studies of protective color change. III. Experiments with fishes both as predators and prey. Proc Natl Acad Sci USA 21:345-353

Swee UB, McCrimmon HR (1966) Reproductive biology of the carp, *Cyprinus carpio* L, in Lake St Lawrence, Ontario. Trans Am Fish Soc 95:372-380

Swerdloff SN (1970) Behavioral observations on Eniwetok damselfishes (Pomacentridae: *Chromis*) with special reference to the spawning of *Chromis caeruleus*. Copeia 1970:371-374

Symons PEK (1974) Territorial behavior of juvenile Atlantic salmon reduces predation by brook trout. Can J Zool 52:677-679

Tautz AF, Groot C (1975) Spawning behavior of chum salmon (*Oncorhynchus keta*) and rainbow trout (*Salmo gairdneri*). J Fish Res Board Can 32:633-642

Tavolga WN (1954) Reproductive behavior in the gobiid fish *Bathygobius soporator*. Bull Am Mus Nat Hist 104:429-459

Tavolga WN (1971) Sound production and detection. In: Hoar WS, Randall DJ (eds) Fish physiology, vol 5. Academic, New York, pp 135-205

Tester AL (1963) The role of olfaction in shark predation. Pac Sci 17:145-170

Thompson WF, Thompson JB (1919) The spawning of the grunion (*Leuresthes tenuis*). Cal Fish Game Fish Bull 3:1-29

Timms AM, Keenleyside MHA (1975) The reproductive behaviour of *Aequidens paraguayensis* (Pisces: Cichlidae). Z Tierpsychol 39:8-23

Tinbergen N (1951) The study of instinct. Clarendon, Oxford, p 228

Tinbergen N (1953) Social behaviour in animals. Methuen, London, p 150

Tinbergen N, Iersel JJA van (1947) "Displacement reactions" in the three-spined stickleback. Behaviour 1:56-63

Trewavas E (1973) On the cichlid fishes of the genus *Pelmatochromis* with proposal of a new genus for *P congicus*; on the relationship between *Pelmatochromis* and *Tilapia* and the recognition of *Sarotherodon* as a distinct genus. Bull Br Mus Nat Hist Zool 25:3-26

Trott LB (1970) Contributions to the biology of carapid fishes (Paracanthopterygii: Gadiformes). Univ Calif Publ Zool 89:1-60

Turner CL (1921) Food of the common Ohio darters. Ohio J Sci 22:41-62

Turner CL (1937) Reproductive cycles and superfetation in poeciliid fishes. Biol Bull 72:145-164

Van Duzer EM (1939) Observations on the breeding habits of the cut-lips minnow, *Exoglossum maxillingua*. Copeia 1939:65-75

Verheijen FJ (1956) Transmission of a flight reaction amongst a school of fish and the underlying sensory mechanisms. Experientia 12:202-204

Verplanck WS (1957) A glossary of some terms used in the objective science of behavior. Psychol Rev 64(6):1-42

Vierke J (1969) Zielstrebige Spuckhandlungen eines Zwergfadenfisches (*Colisa lalia*). Bonn Zool Beitr 20:408-415

Vierke J (1973) Das Wasserspucken der Arten der Gattung *Colisa* (Pisces: Anabantidae). Bonn Zool Beitr 24:62-104

Vierke J (1975) Beiträge zur Ethologie und Phylogenie der Familie Belontiidae (Anabantoidei, Pisces). Z Tierpsychol 38:163-199

Vine I (1971) Risk of visual detection and pursuit by a predator and the selective advantage of flocking behaviour. J Theor Biol 30:405-422

Vine PJ (1974) Effects of algal grazing and aggressive behaviour of the fishes *Pomacentrus lividus* and *Acanthurus sohal* on coral-reef ecology. Mar Biol 24:131-136

Walford LA (1958) Living resources of the sea. Ronald, New York, p 321

Ward JA, Barlow GW (1967) The maturation and regulation of glancing off the parents by young orange chromides (*Etroplus maculatus*: Pisces-Cichlidae). Behaviour 29:1-56

Ward JA, Wyman RL (1977) Ethology and ecology of cichlid fishes of the genus *Etroplus* in Sri Lanka: preliminary findings. Environ Biol Fish 2:137-145

Warner RR, Robertson DR, Leigh EG (1975) Sex change and sexual selection. Science 190:633-638

Webb PW (1973) Kinematics of pectoral fin propulsion in *Cymatogaster aggregata*. J Exp Biol 59:697-710

Webb PW (1975a) Hydrodynamics and energetics of fish propulsion. Bull Fish Res Board Can 190:1-158

Webb PW (1975b) Efficiency of pectoral-fin propulsion of *Cymatogaster aggregata*. In: Wu TYT, Brokaw CJ, Brennan C (eds) Swimming and flying in nature, vol 2. Plenum, New York, pp 573-584

Weihs D (1973) Hydromechanics of fish schooling. Nature (London) 241:290-291

Weihs D (1975) Some hydrodynamical aspects of fish schooling. In: Wu TYT, Brokaw CJ, Brennan C (eds) Swimming and flying in nature, vol 2. Plenum, New York, pp 703-718

Weisel GF, Newman HW (1951) Breeding habits, development and early life history of *Richardsonius balteatus*, a northwestern minnow. Copeia 1951:187-194

Weitzman SH (1954) The osteology and the relationships of the South American characid fishes of the subfamily Gasteropelecinae. Stanford Ichthyol Bull 4:213-263

Werner EE, Hall DJ (1974) Optimal foraging and the size selection of prey by the bluegill sunfish (*Lepomis macrochirus*). Ecology 55:1042-1052

Wickett PW (1954) The oxygen supply to salmon eggs in spawning beds. J Fish Res Board Can 11:933-953

Wickett PW (1959) Observations on adult pink salmon behavior. Fish Res Board Can Pac Biol Stn Prog Rep 113:6-7

Wickler W (1955) Das Fortpflanzungsverhalten der Keilfleckbarbe, *Rasbora heteromorpha* Duncker. Z Tierpsychol 12:220-228

Wickler W (1962a) Ei-Attrappen und Maulbrüten bei afrikanischen Cichliden. Z Tierpsychol 19:129-164

Wickler W (1962b) "Egg-dummies" as natural releasers in mouth-breeding Cichlids. Nature (London) 194:1092-1093

Wickler W (1965a) Signal value of the genital tassel in the male *Tilapia macrochir* Blgr (Pisces: Cichlidae). Nature (London) 208:595-596

Wickler W (1965b) Neue Varianten des Fortpflanzungsverhaltens afrikanischer Cichliden (Pisces: Perciformes). Naturwissenschaften 52:219

Wickler W (1966a) Ein Augen fressender Buntbarsch. Nat Mus Frankf 96:311-315

Wickler W (1966b) Über die biologische Bedeutung des Genitalanhanges der männlichen *Tilapia macrochir*. Senckenberg Biol 47:419-427

Wickler W (1967) Vergleich des Ablaichverhaltens einiger paarbildender sowie nicht-paarbildender Pomacentriden und Cichliden (Pisces: Perciformes). Z Tierpsychol 24:457-470

Wickler W (1968) Mimicry in plants and animals. McGraw-Hill, New York, p 255

Wiepkema PR (1961) An ethological analysis of the reproductive behaviour of the bitterling (*Rhodeus amarus* Bloch). Arch Néerl Zool 14:103-199

Williams GC (1964) Measurement of consociation among fishes and comments on the evolution of schooling. Pap Mus Mich State Univ Biol Ser 2:351-383

Williams GC (1975) Sex and evolution. Princeton Univ Press, Princeton NJ, p 200

Williams NJ (1972) On the ontogeny of behaviour of the cichlid fish *Cichlasoma nigrofasciatum* (Günther). Doctoral Thesis, Univ Groningen, Holland, p 111

Wilson EO (1971) Competitive and aggressive behavior. In: Eisenberg JF, Dillon W (eds) Man and beast: comparative social behavior. Smithson Inst, Washington, pp 183-217

Wilson EO (1975) Sociobiology, the new synthesis. Belknap, Harvard Univ Press, Cambridge, p 697

Wilz KJ (1970) Causal and functional analysis of dorsal pricking and nest activity in the courtship of the three-spined stickleback *Gasterosteus aculeatus*. Anim Behav 18:115-124

Wilz KJ (1973) Quantitative differences in the courtship of two populations of three-spined sticklebacks, *Gasterosteus aculeatus*. Z Tierpsychol 33:141-146

Winn HE (1958a) Observations on the reproductive habits of darters (Pisces: Percidae). Am Midl Nat 59:190-212

Winn HE (1958b) Comparative reproductive behavior and ecology of fourteen species of darters (Pisces: Percidae). Ecol Monogr 28:155-191

Winn HE, Bardach JE (1960) Some aspects of the comparative biology of parrotfishes at Bermuda. Zoologica 45:29-34

Wong B, Ward FJ (1972) Size selection of *Daphnia pulicaria* by yellow perch (*Perca flavescens*) fry in West Blue Lake, Manitoba. J Fish Res Board Can 29:1761-1764

Wootton RJ (1971) A note on the nest-raiding behavior of male sticklebacks. Can J Zool 49:960-962

Wootton RJ (1972) The behaviour of the male three-spined stickleback in a natural situation: a quantitative description. Behaviour 41:232-241

Wootton RJ (1976) The biology of sticklebacks. Academic, London, p 387

Wu TYT, Brokaw CJ, Brennan C (eds) (1975) Swimming and flying in nature, vol 2. Plenum, New York, p 1005

Wyman RL, Ward JA (1973) The development of behavior in the cichlid fish *Etroplus maculatus* (Bloch). Z Tierpsychol 33:461-491

197

# Systematic Index

199

200

201

204

# Subject Index

# Zoophysiology and Ecology

Coordination Editor: D. S. Farner
Editors: W. S. Hoar, B. Hoelldobler,
H. Langer, M. Lindauer

Springer-Verlag
Berlin
Heidelberg
New York

*A Springer journal*

# Behavioral Ecology and Sociobiology

Managing Editor: H. Markl
in cooperation with a distinguished advisory
board

*Behavioral Ecology and Sociobiology* was founded by Springer-Verlag in 1976 as an international journal. Drawing on a philosophy
developed and nurtured for more than half a
century in the original *Zeitschrift für Vergleichende Physiologie* (now *Journal of Comparative
Physiology*), it presents original articles and short
communications dealing with the experimental
analysis of animal behavior on an individual
level and in population. Special emphasis is
given to the functions, mechanisms, and evolution of ecological adaptations of behavior.
Specific areas covered include:

- orientation in space and time
- communication and oll other forms of social
  and interspecific behavior
- origins and mechanisms of behavior preferences and aversions, e.g., with respect to food,
  locality, and social partners
- behavioral mechanisms of competition and
  resource partitioning
- population physiology
- evolutionary theory of social behavior.

*Behavioral Ecology and Sociobiology* is designed
to serve as a link between researchers and students in a variety of disciplines.

**Subscription Information** upon request.

Springer-Verlag
Berlin
Heidelberg
New York

# Journal of Comparative Physiology · A+B

Founded in 1924 as
Zeitschrift für Vergleichende Physiologie
by K. von Frisch and A. Kühn

**A.** Sensory, Neural, and Behavioral Physiology

**Editorial Board:** H. Autrum, R. R. Capranica,
K. von Frisch, G. A. Horridge, M. Lindauer,
C. L. Prosser

**Advisory Board:** H. Atwood, S. Daan,
W. H. Fahrenbach, B. Hölldobler, Y. Katsuki,
M. Konishi, M. F. Land, M. S. Laverack,
H. C. Lüttgau, H. Markl, A. Michelsen, D. Otto-
son, F. Papi, C. S. Pittendrigh, W. Precht,
J. D. Pye, A. Roth, H. F. Rowell, D. G. Stavenga,
R. Wehner, J. J. Wine

**B.** Biochemical, Systemic, and Environmental
Physiology

**Editorial Board:** K. Johansen, B. Linzen,
W. T. W. Potts, C. L. Prosser

**Advisory Board:** G. A. Bartholomew, H. Bern,
P. J. Butler, Th. Eisner, D. H. Evans, S. Nilsson,
O. Randall, R. B. Reeves, G. H. Satchell,
T. J. Shuttleworth, G. Somero, K. Urich,
S. Utida, G. R. Wyatt, E. Zebe

The Journal of Comparative Physiology
publishes original articles in the field of animal
physiology. In view of the increasing number
of papers and the high degree of scientific
specialization the journal is published in two
sections.

**A. Sensory, Neural, and Behavioral Physiology**
Physiological Basis of Behavior; Sensory
Physiology; Neural Physiology; Orientation,
Communication; Locomotion; Hormonal
Control of Behavior

**B. Biochemical, Systematic and Environmental
Physiology**
Comparative Aspects of Metabolism and Enzy-
mology; Metabolic Regulation, Respiration
and Gas Transport; Physiology of Body Fluids;
Circulation; Temperature Relations; Muscular
Physiology

**Subscription Information** upon request.